AF587663

MICROBIOLOGY RESEARCH ADVANCES

NUCLEAR RECEPTORS

Microbiology Research Advances

Additional books in this series can be found on Nova's website under the Series tab.

Additional E-books in this series can be found on Nova's website under the E-book tab.

Cell Biology Research Progress

Additional books in this series can be found on Nova's website under the Series tab.

Additional E-books in this series can be found on Nova's website under the E-book tab.

MICROBIOLOGY RESEARCH ADVANCES

NUCLEAR RECEPTORS

MARGARET K. BATES
AND
REGINA M. KERR
EDITORS

Nova Science Publishers, Inc.
New York

For permission to use material from this book please contact us:
Telephone 631-231-7269; Fax 631-231-8175
Web Site: http://www.novapublishers.com

Library of Congress Cataloging-in-Publication Data

Nuclear receptors / editors, Margaret K. Bates and Regina M. Kerr.
p. ; cm.
Includes bibliographical references and index.
ISBN 978-1-61209-980-4 (hardcover)
1. Nuclear receptor (Biochemistry) I. Bates, Margaret K. II. Kerr, Regina M. III. Title.
[DNLM: 1. Receptors, Cytoplasmic and Nuclear. QU 58.5]
QH603.C43N838 2011
572.8'845--dc22 2011007118

Published by Nova Science Publishers, Inc. † New York

CONTENTS

PREFACE

Nuclear receptors are involved in various aspects of intracellular signal transduction within a range of tissues and play an important role as regulators in numerous essential biological functions. In this new book, the authors present topical research in the study of nuclear receptors, including glucocorticoid receptor signaling in cardiovascular disease; steroid receptor coactivators and endocrine treatment in breast cancer; the effects of nitric oxide on nuclear receptors as a tool for studying gene regulation; vitamin K as a ligand of steroid and xenobiotic receptors and androgen receptors and prostate cancer.

Chapter 1 – Glucocorticoid administration has been considered for the reduction of atherosclerosis and restenosis following percutaneous coronary intervention. Excess of glucocorticoids (GCs), due to either chronic exogenous treatment or to endogenous hypersecretion of cortisol (such as depression, systemic inflammation and Cushing disease) has been associated with increased cardiovascular risk. The GC receptor is the cornerstone molecule mediating GCs' cellular effects. Myocyte sensitivity to GCs depends on intracellular, pre-receptor metabolism of active cortisol to inactive cortisone by the enzyme 11β-hydroxy steroid dehydrogenase. The molecular basis of GC-induced cardiovascular effects such as atherosclerosis and hypertension remain elusive. Toll-like receptors (TLRs) represent the first line of host defense against microbial infection and play a significant role in both innate and adaptive immunity by recognition of exogenous pathogen-associated molecular patterns and endogenous ligands. The TLR and GR signaling pathways interact and modulate the inflammatory response. Innate immune and inflammatory pathways have been implicated in cardiac dysfunction after global myocardial ischemia. There are 10 TLRs identified to date and they bind to a variety of pathogenic agents such as lipopeptide (TLR2) and lipopolysaccharide (TLR4) by molecular pattern recognition. TLR4 is expressed in myocardial cells and increased expression of this receptor is observed in cardiac myocytes from human and animal models of ischemic cardiac injury. Stimulation of TLRs leads to the activation of various downstream transcription factors in particular nuclear factor (NF)-κB and the production of inflammatory cytokines in myocardial cells. These cytokines in turn activate TLRs in a positive feedback mechanism that sustains inflammation thereby contributing to disease progression. S100B, a member of the S100 family of calcium-binding proteins is induced by adrenergic stimulation in myocardial cells following ischemic cardiac injury and depending on the concentration achieved, via receptor for advanced ligation endproducts (RAGE) ligation, results in activation of NF-κB and trophic or apoptotic cell responses. Preliminary data from hypoxic myocardial cells demonstrate a possible association between TLR4 and S100B. The common signaling pathway involves the activation of NF-κB leading to GC receptor activation and transrepression of target genes. In this chapter we provide a comprehensive review of GC receptor signaling and its convergence with TLR4- and RAGE-S100B dependent signaling pathways as it relates to inflammation and the progression of cardiovascular disease.

Chapter 2 – Breast cancer is the most frequent malignancy in women in the Western world. Hormone receptor positive breast cancers are managed with endocrine treatment in which the estrogen receptor (ER) is blocked using a selective estrogen receptor modulator (SERM) such as tamoxifen or by targeting the estrogen synthesis using aromatase inhibitors (AIs). Nuclear receptor coactivators have been pointed out as the main determinants of tissue-, cell- and promoter specific effects of tamoxifen, and they are important regulators of ER mediated gene transcription under estrogen deprivation induced by aromatase inhibition.

The steroid receptor coactivator (SRC) family comprises SRC-1, SRC-2/TIF-2 and SRC-3/AIB1. Typically they enhance the transcriptional activity of ligand-bound ER by binding to the activation function-2 (AF-2) pocket, and recruit the basal transcription factors and chromatin-remodeling complex, acetyltransferase proteins, methyltransferases and ubiquitin ligases. 4-hydroxytamoxifen works as an ER antagonist by binding to the nuclear receptor and inducing a displacement of helix 12 that blocks binding of coactivators and favors corepressor recruitment. However, high levels of coactivators relative to corepressors may force ER into an active structural conformation where 4-hydroxytamoxifen leads to ER agonistic effects via AF-

1 by indirect binding to DNA and recruitment of coactivators. While SRC-1 and SRC-2/TIF-2 are expressed in normal and malignant breast tissue, SRC-3/AIB1 predominates with overexpression in >30% and gene amplification in 5 – 10% of breast cancers. Our recent studies in human breast cancer have shown that treatment with tamoxifen or AIs enhances gene expression of the SRCs. Others have reported that high levels of SRC-1 are associated with nodal involvement and resistance to endocrine treatment. SRC-3/AIB1 and the growth factor receptor HER-2/neu are often coexpressed in breast cancers, and poor response to tamoxifen treatment and reduced disease-free survival are found when tumors overexpress SRC-1 or SRC-3/AIB1 together with HER-2/neu. The SRCs are regulated by post-translational modifications by for instance mitogen activated protein kinases (MAPKs) which operate downstream of HER-2/neu and stabilize and functionally activate SRC proteins, a mechanism which could contribute not only to tamoxifen resistance, but also to estrogen hypersensitivity and resistance to AIs. In summary, steroid receptor coactivators are crucial in ER regulated gene transcription. Accumulated evidence points to an association between coactivator levels, effect of and response to endocrine treatment and long-term outcome in human breast cancer. The SRCs are involved in crosstalk between ER and growth factor pathways that are activated in breast cancer, making coactivators important in breast cancer development and interesting as potential therapeutic targets.

Chapter 3 – Nuclear receptors are involved in various aspects of intracellular signal transduction on a range of tissue and play an important role as regulators in numerous essential biological functions. In the thymus, these nuclear receptors also participate in positive or negative selection during T cell development.

In particular, the glucocorticoid receptor (Gr) and Nur77 play central role s in apoptosis induction mediated by the T cell receptor (TCR) in mature thymocytes or glucocorticoids (GCs) in immature thymocytes, respectively. Recently, we demonstrated that death-associated protein 3 (DAP3) was critical for TCR-mediated induction of apoptosis downstream of Nur77 in immature thymocytes. The DAP3 is an evolutionarily conserved GTP binding protein that plays a number of roles in normal mitochondrial physiology and in apoptosis induced via the tumor necrosis factor (TNF) family of death receptors. This chapter reviews recent studies of the signal transduction mediated by Gr and Nur77 in thymocyte development, focusing on signaling molecules, such as DAP3, involved in the signaling pathways of Gr or Nur77. Briefly, discussion which have attracted attention are summarized as follows:

(1). signaling molecules interacting with Gr or Nur77, (2). the functional role of Gr or Nur77 in subcellular localization, (3). the function of genes subject to induction by Gr or Nur77, (4). crosstalk and its physiological importance in the signal transduction mediated by Gr and Nur77.

Chapter 4 – Nitric oxide (NO), a free radical gas, is an omnipresent intercellular messenger in all vertebrates. Originally described as a cardiovascular signal molecule NO elicites a variety of physiological functionslike muscle contractility, platelet aggregation, metabolism, neuronal activity, and immune responses in a broad range of tissues.

The molecule originates from the action of nitric oxide synthases (NOS) which are either induced (iNOS) or constitutivcly expressed (eNOS, nNOS). The underlying mechanisms of NO action are primarily an elevation of guanosine 3',5'-cyclic monophosphate due to the stimulation of soluble guanylyl cyclase, inhibition of mitochondria respiration and nitrosylation of proteins.

In vitro, NO has been shown to modulate the activity of a variety of nuclear receptors (NR) of the steroid receptor superfamily like the estrogen (ER) or the androgen receptor (AR).

Nuclear receptors are transcription factors characterized by a ligand binding domain, a highly conserved DNA binding domain consisting of two Cys4-type zinc fingers and a transactivation domain, possessing one of the two activation function domains.

Upon ligand binding, NR predominantly act via binding to so called hormone response elements on the DNA thus regulating gene expression. Essential features which regulate the activity and function of NR are receptor concentration, post translational modifications like phosphorylation or acetylation which trigger receptor dimerization, nucleocytoplasmic shuttling or binding of comodulators.

Despite their different modes of action, the signaling pathways of nitric oxide and NR interfere a manifold. Estrogen and progesterone are known to up-regulate NO synthesis whereas glucocorticoids and progesterone decrease NO bioavailability. Due to its unique physicochemical properties (high reactivity of NO-radicals, short half-life, excellent membrane permeability) the NO-molecule is also able to directly interact with nuclear receptors thereby blocking their activity.

In the chapter, the authors will summarize the so far known effects of NO on nuclear receptors and demonstrate its potential use as a tool for studying gene regulation. In addition they will discuss the physiological relevance of NO/NR interaction.

Chapter 5 – Vitamin K is a fat-soluble vitamin essential for blood coagulation. Natural vitamin K includes vitamin K1 (phylloquinone), present mainly in vegetables, and vitamin K2 (menaquinone), which is synthesized by microorganisms and is present in food such as natto (fermented soy beans). Vitamin K1 is converted to vitamin K2, the functionally active form, in the body.

Vitamin K was shown to play an essential role in the hepatocytes by maintaining the activity of coagulation factors II, VII, IX, and X and of anticoagulants, protein C, and protein S. Recently, extrahepatic actions of vitamin K have also been reported.

The administration of vitamin K was shown to prevent bone fracture, and this led to its clinical application in cases of osteoporosis in East Asian countries. Epidemiological studies have shown that the lack of vitamin K causes osteoarthritis and imposes a risk of coronary artery disease. Some clinical studies have suggested its antitumor activity in cases of hepatocellular carcinoma and other cancers.

Vitamin K is a natural and safe nutrient. The novel effect of vitamin K that involves the nuclear receptor SXR/PXR has been elucidated. It can be presumed that this novel effect exists in other organs and tissues where SXR/PXR expression is detected, like in the bone and in the liver.

Further investigation on this mechanism is warranted to understand the precise function of vitamin K and the pathology of the concerning diseases, and to discover new therapeutic approaches for these diseases.

Chapter 6 – Nuclear receptors are a class of proteins that have the ability to directly bind to DNA and regulate gene expression, and these receptors are classified as transcription factors. This report focuses on a new function of AR (androgen receptor). Androgen receptors (ARs) belong to the steroid receptor family and play an essential role in the generation and development of the prostate. Androgen receptors have similar conserved domains that are composed of an NTD (N-terminal domain), a DBD (DNA-binding domain), and an LBD (ligand-binding domain). The NTD works stabilize bound androgen and the AR-LBD mediates the interaction between AR and other proteins, which include Hsps (heat-shock proteins). In the absence of androgen, AR remains in the cytoplasm in an inactive form. After AR binds to androgens, activated AR can bind with other signal molecules and form functional complexes. Then, the complex translocates to the nucleus and regulates the gene expression for androgen regulated genes. Recently, some research has shown that AR can interact with DDC (L-dopa decarboxylase), a key molecule for serotonin (5-HT) synthesizing. Serotonin is a well known neurotransmitter but has been mentioned in the relationship with the generation of the prostate. Then, we introduce here that AR can regulate prostate cancer progression via the serotonin synthesis process.

A suggested rewrite of the previous sentence, placed here to avoid ambiguity: This chapter suggests that AR may play a role in regulating the progress of prostate cancer via the serotonin synthesis process.

Chapter 7 – Plants are immobile and, therefore, confronted with a variety of environmental stresses. To adapt to these harmful conditions, plants have developed effective and complex stress signal transduction pathways between the nucleus and cytoplasm for response to and survival from stress conditions. In eukaryotic cells, the nuclear envelope separates the cytoplasm from the nucleus, and the nuclear pore complex (NPC) is the gateway for signal molecules trafficking across the nuclear envelope. Small molecules utilize passive diffusion to pass through the NPC; however, the efficient and directed translocation of macromolecules requires nuclear transport receptors to facilitate the passage through the NPC. In this regard, importinβ-like nuclear transport receptors are the main receptors for nuclear transport in Arabidopsis. Some of these receptors act as nuclear import receptors (importins) and some as nuclear export receptors (exportins). In addition, many stress responses in plants are controlled by the nucleocytoplasmic partitioning of regulatory molecules between the cytoplasm and nucleoplasm, such as light, temperature, and responses to cytokinin-signal transduction and pathogen infection. Thus, importins and exportins are very important for plant growth, development, and stress response.

Chapter 8 – Peroxisome proliferator-activated receptors (PPARs) are ligand-activated nuclear receptors that regulate gene expression and are modulated by interaction with corepressors and coactivators. Natural and synthetic ligands promote heterodimerization of PPARs with the retinoid-X-receptor (RXR), facilitating their binding to consensus DNA sequences on target genes. To date, three subtypes of PPAR have been identified – α, δ, and γ; with the γ subtype consisting of two distinct functional isoforms, γ1 and γ2. Although structurally similar, the PPAR subtypes have specific tissue distribution and functions. PPARs regulate multiple cellular functions, such as cell proliferation, the immune response and lipid metabolism and therefore their ligands have been investigated for their potential use in various clinical settings. For example, the PPARγ ligands comprising the thiazolidinedione class of drugs have been used for the management of type 2diabetes. The potential use of PPAR ligands in autoimmune diseases is also being investigated whereas the specific role of PPARs in tumor development is still controversial. Despite some drawbacks, PPARs still remain as potential therapeutic targets for various conditions and currently dual and pan agonists are being investigated for this purpose. Taken together, the role of PPARs in various cellular processes and disease pathogenesis still requires further investigation and continues to be an exciting field of research. This review will attempt to provide examples of some of the recent findings in these areas of research, highlighting the mechanisms of action and the potential use of PPAR agonists as well as the challenges that still need to be addressed.

In: Nuclear Receptors
Editors: M. K. Bates and R. M. Kerr

ISBN: 978-1-61209-980-4

Chapter 1

GLUCOCORTICOID RECEPTOR SIGNALING AND ITS POTENTIAL CONVERGENCE WITH TOLL-LIKE RECEPTOR (TLR)-4 AND RECEPTOR FOR ADVANCED GLYCATION END-PRODUCTS (RAGE) SIGNALING PATHWAYS IN CARDIOVASCULAR DISEASE

Vasileios Salpeas[1,3], James N. Tsoporis[2], Shehla Izhar[2], Thomas G. Parker[2], Ekaterini S. Bei[3], Paraskevi Moutsatsou[3] and Ioannis K. Rizos[1]

[1]2nd Academic Department of Cardiology, Attikon University Hospital, University of Athens Medical School

[2]Division of Cardiology, Department of Medicine, Keenen Research Centre, Li Ka Shing Knowledge Institute, St. Michael's Hospital, University of Toronto, Toronto, Ontario, Canada

[3]Laboratory of Biological Chemistry, Medical School, University of Athens, Athens, Greece

ABSTRACT

Glucocorticoid administration has been considered for the reduction of atherosclerosis and restenosis following percutaneous coronary intervention.

Excess of glucocorticoids (GCs), due to either chronic exogenous treatment or to endogenous hypersecretion of cortisol (such as depression, systemic inflammation and Cushing disease) has been associated with increased cardiovascular risk. The GC receptor is the cornerstone molecule mediating GCs' cellular effects. Myocyte sensitivity to GCs depends on intracellular, pre-receptor metabolism of active cortisol to inactive cortisone by the enzyme 11β-hydroxy steroid dehydrogenase. The molecular basis of GC-induced cardiovascular effects such as atherosclerosis and hypertension remain elusive. Toll-like receptors (TLRs) represent the first line of host defense against microbial infection and play a significant role in both innate and adaptive immunity by recognition of exogenous pathogen-associated molecular patterns and endogenous ligands. The TLR and GR signaling pathways interact and modulate the inflammatory response. Innate immune and inflammatory pathways have been implicated in cardiac dysfunction after global myocardial ischemia. There are 10 TLRs identified to date and they bind to a variety of pathogenic agents such as lipopeptide (TLR2) and lipopolysaccharide (TLR4) by molecular pattern recognition. TLR4 is expressed in myocardial cells and increased expression of this receptor is observed in cardiac myocytes from human and animal models of ischemic cardiac injury. Stimulation of TLRs leads to the activation of various downstream transcription factors in particular nuclear factor (NF)-κB and the production of inflammatory cytokines in myocardial cells. These cytokines in turn activate TLRs in a positive feedback mechanism that sustains inflammation thereby contributing to disease progression. S100B, a member of the S100 family of calcium-binding proteins is induced by adrenergic stimulation in myocardial cells following ischemic cardiac injury and depending on the concentration achieved, via receptor for advanced ligation endproducts (RAGE) ligation, results in activation of NF-κB and trophic or apoptotic cell responses. Preliminary data from hypoxic myocardial cells demonstrate a possible association between TLR4 and S100B. The common signaling pathway involves the activation of NF-κB leading to GC receptor activation and transrepression of target genes. In this chapter we provide a comprehensive review of GC receptor signaling and its convergence with TLR4- and RAGE-S100B dependent signaling pathways as it relates to inflammation and the progression of cardiovascular disease.

S100 Proteins and Rage Signaling

S100 Proteins Structure and Function

S100 proteins entail a multigenic family of calcium binding proteins of the EF- hand type (helix E-loop-helix F). These proteins are called S100 because of their solubility in a 100% -saturated solution with ammonium sulphate at neutral pH. They are small acidic proteins, 10-12KDa and contain two distinct EF-hands, 4 α-helical segments, a central hinge region of variable length and the N-and C-terminal variable domains. Twenty five members of this family have been identified (Donato et al., 2009). Of these, 21 (S100A1-S100A18, trichohylin, filagrin and repetin) have genes clustered on a 1.6-Mbp segment of human chromosome 1 (1q21) while other members are found at chromosome loci 4q16 (S100P), 5q14 (S100Z), 21q22 (S100B), and Xp22 (S100G) (Engelkamp et al., 1993). S100 proteins are widely expressed in a variety of cell types and tissues. For example, S100A1 and S100A2 are found in the cytoplasm and nucleus of smooth-muscle cells of skeletal muscle, respectively (Donato 2001), S100P is located in the cytoplasm of placental tissue (Becker et al., 1992; Emoto et al., 1992) and S100B in cytoplasm of astrocytes of nervous system (Kligman and Marshak 1985). S100 proteins do not exhibit intrinsic catalytic activity but are calcium sensor proteins and through interaction with several intracellular effector proteins they contribute to the regulation of a broad range of functions such as contraction, motility, cell growth and differentiation, cell cycle progression, organization of membrane-associated cytoskeleton elements, cell survival, apoptosis, protein phosphorylation and secretion (Donato 1999, 2001; Donato et al 2009). For example, S100B regulates the cytoskeletal dynamics through disassembly of tubulin filaments, type III intermediate filaments (Donato et al., 2009) and binding to fibrillary proteins such as CapZ (Kilby et al., 1997) or inhibiting GFAP phosphorylation when stimulated by cAMP or calcium/calmodulin (Frizzo et al., 2004). To modulate these types of activities S100 proteins undergo conformational changes (Ikura 1996). Upon calcium binding the helices of S100 proteins rearrange, revealing a hydrophobic cleft, which forms the target protein binding site (Rustandi et al., 2000). Although target binding of S100 proteins is calcium-dependent, calcium independent interactions have been reported (Santamaria-Kisiel et al., 2006). Enzymes are the most common calcium independent target binding for the S100 proteins. S100B for example, interacts in a calcium-dependent manner with the cytoplasmic domain of myelin-associated glycoprotein and inhibits its phosphorylation by protein kinase (Kursula et al., 2000). It is also implicated in tau protein (Baudier and Cole 1988) and p53 phosphorylation (Markowitz et al., 2005), inhibition of Ndr kinase activity (Millward etal., 1998), and regulation of the activity of the GTPase Rac1 and Cdc42 effector IQGAP1 (Mbele et al., 2002). The most significant calcium-independent interactions of S100 proteins are their ability to bind to each other. They form homodimers, but heterodimerization adds to the complexity of this multiprotein family. Each subunit consists of two helix-loop-helix motifs connected by a central linker or so-called hinge region. The C-terminal canonical EF-hand motif is composed of 12 amino acids, whereas the N-terminal S100-specific EF-hand comprises 14 residues (Engelkamp et al., 1993; Heizmann et al., 1998). In addition to intracellular activities, some S100 proteins (e.g. S100B, S100A1, S100A4, S100A8, S100A9) are secreted by the cell and exhibit extracellular functions (Donato 2003). The S100A8/A9 heterodimer is secreted by a novel secretion pathway that depends on an intact microtubule network and acts as a chemotactic molecule in inflammation (Kerkhoff et al., 1998; Newton and Hogg 1998). S100B is secreted by a number of cell types (e.g. astrocytes, glial cells) (Van Eldik and Zimmer 1987). Astrocytes and glial cells secrete S100B, by a complex system involving alterations in intracellular calcium concentration (Van Eldik and Zimmer 1987). S100B after secretion, or simply leakage from damaged cells, accumulates in the extracellular space and/or enters the blood stream and cerebrospinal fluid (Peskind et al., 2001; Portela et al., 2002). The extracellular effects of some S100 proteins require binding to the receptor for advanced glycosylation end products (RAGE) (Schmidt et al., 2001; Bierhaus et al., 2005; Leclerc et al., 2009; Heizmann et al., 2007). RAGE is a member of the immunoglobulin family of cell surface molecules recognizing multiple ligands including AGE, amphoterin, amyloid-β-peptide and β-fibrils, high mobility group box-1 (HMGB1) S100A12, S100A6, and S100B (Donato 2007). The 45-kDa receptor protein consists of 403 amino acids with an extracellular domain (1 variable and 2 constant Ig domains with disulfide bridges), a single trans-membrane region, and a short cytosolic tail that triggers signal transduction (Leclerc et al., 2007). RAGE ligands show selective binding to RAGE. S100B tetramer induces receptor dimerization by binding to RAGE (Ostendorp et al., 2007). S100B binds to domains V and CI where as the RAGE ligand S100A6 binds to domains CI and CII (Leclerc et al., 2007). Ligand binding to RAGE results in increased expression of activated NF-κBp65 (Donato 2003, 2007; Donato et al., 2009). Interestingly, the presence of NF-κB binding sites in the RAGE promoter creates a positive feedback loop resulting in increased RAGE expression (Donato 2003, 2007). The NF-κB family of transcription factors contain 5 family members that function as hetero- or homodimers. The dimmers are sequestered in the cytoplasm in an inactive form by IκB. NF-κB is activated when IκB is phosphorylated and subsequently degraded, resulting in the translocation of NF-κB to the nucleus (Karin and Ben-Neriah 2000). The action of S100B is strongly dependent on its extracellular concentration. At nanomolar quantities it has trophic effects however at micromolar concentrations it promotes apoptosis (Hu and Van Eldik 1996; Huttunen et al., 2000). Micromolar extracellular levels are detected after brain injury or in neurodegenerative disorders like Down's Syndrome, Alzheimer disease or encephalitis (Van Eldik and Griffin 1994; Griffin et al., 1998). Trophic and toxic effects of extracellular S100B are mediated in the brain by RAGE (Huttunen et al., 2000). In addition, to playing a major role in brain physiology (Donato et al., 2009), S100B has been implicated in cardiovascular development (Schaub and Heizmann 2008) and is considered a biochemical marker for brain injuries after bypass graft surgery (Anderson et al., 1999) and dilated cardiomyopathy (Mazzini et al., 2007).

Cardiovascular Effects of Intracellular S100B

Experimental evidence suggests that S100B acts as an intrinsic negative regulator of the myocardial hypertrophic response (Tsoporis et al., 1998, 2005, 2009). Negative modulators of the hypertrophic response are essential to maintain a balance between compensatory hypertrophy and unchecked progression. S100B not normally expressed in the myocardium, is induced in the peri-infarct region of the human heart after myocardial infarction (Tsoporis et al., 1998) and in rat heart commencing at day 7 following myocardial infarction as a result of experimental coronary artery ligation (Tsoporis et al., 1997). In cultured neonatal rat cardiac myocytes, transfection of an expression vector encoding the human S100B protein, inhibits the α_1-adrenergic induction of embryonic β-myosin heavy chain (MHC), the skeletal isoform of α actin (skACT), and atrial natriuretic factor (ANF) by interrupting the PKC signaling pathway in a calcium dependent manner (Tsoporis et al., 1997). The molecular mechanisms by which S100B modulates the hypertrophic phenotype remain to be defined. To provide a physiologic model of S100B overexpression effects, transgenic mice were created that contained multiple copies of the human gene under the control of its own promoter. These animals demonstrate normal cardiac structure, and neuronal, but no basal cardiac expression of the transgene. In S100B transgenic mice, after chronic α_1-adrenergic agonist infusion, S100B is detected in the heart and increased in the vasculature and the myocyte hypertrophy and arterial smooth muscle cell proliferation normally evoked in the heart and vasculature respectively in response to α_1-adrenergic stimulation in wild-type mice were abrogated (Tsoporis et al., 2009). In S100B knockout mice, α_1-adrenergic agonist infusion provoked a potentiated myocyte hypertrophic response and augmented arterial smooth muscle cell proliferation. Furthermore, in S100B knockout mice, both the acute and chronic increase in blood pressure in response to α_1-adrenergic agonist infusion was attenuated compared with wild-type mice (Tsoporis et al., 2009). To determine whether this inhibition is generalizable to other hypertrophic stimuli, S100B transgenic and knock-out animals were subjected to descending aortic banding to produce pressure-overload. Aortic banding for 35 days induced a hypertrophic response in wild-type mice, no hypertrophy in S100B transgenic mice and excessive hypertrophy in S100B knock-out mice. Similarly, 35 days after experimental myocardial infarction, the S100B knockout mice mounted an augmented hypertrophic response compared to wild-type mice (Tsoporis et al., 2005). Fetal gene expression was induced to a greater magnitude in S100B knockout mice compared to wild-type mice. The S100B transgenic mice did not develop the hypertrophic phenotype but demonstrated increased apoptosis in the peri-infarct region compared to wild-type and S100B knockout mice. These studies in S100B transgenic and knockout mice complement the culture data and support the hypothesis that intracellular S100B acts both as an intrinsic negative regulator of hypertrophy and an apoptotic agent. Intracellular S100B may modulate the apoptotic response of post-infarct myocytes by activating a transcriptionally inducible form of nitric oxide synthase (iNOS) and production of nitric oxide (NO) (Razavi et al., 2005) as has been described for astrocytes (Hu and Van Eldik 1996). Thus, NO could be an intermediate pathway in the induction of apoptosis by intracellular S100B.

Cardiovascular Effects of Extracellular S100B via RAGE Signaling

Increasing evidence suggests that S100B plays a role in the regulation of apoptosis in post-MI myocardium by an extracellular mechanism after cellular release from damaged myocytes and interaction with RAGE (Tsoporis et al., 2010). Exogenously administered S100B to neonatal rat cultures induced apoptosis in a dose-dependent manner beginning at 0.05 μmol/L, a local or regional concentration that may be achieved in the peri-infarct myocardium (Tsoporis et al., 2005). Similarly, S100B at dose $\geq$0.05 μmol/L induced neuronal cell death (Iuvone et al., 2007). Myocyte apoptosis is accompanied by cytochrome C release from mytochondria to cytoplasm, increased expression and activity of pro-apoptotic caspase-3, decreased expression of anti-apoptotic Bcl-2 and phosphorylation of ERK1/2 and p53 (Delphin et al., 1999; Goncalves et al., 2000; Tsoporis et al., 2010). Transfection of a full-length cDNA of RAGE or a dominant-negative mutant of RAGE resulted in increased or attenuated S100B-induced myocyte apoptosis respectively, implicating RAGE dependence. Inhibition of MEK signaling and/or overexpression of a dominant negative p53 inhibits S100B-induced myocyte apoptosis. This implies that RAGE activation by S100B increases MEK MAPK kinase signaling, p53 phosphorylation at serine 15 and p53-dependent myocyte apoptosis.

Toll-Like Receptors

Toll-Like Receptors and S100B

Although RAGE is an important S100B receptor, RAGE engagement may not be the sole means whereby S100B brings about its effects on target cells in particular cardiovascular cells. We now provide evidence that toll-like receptor 4 (TLR4) is upregulated in cardiac myocytes following myocardial infarction as previously described (Frantz et al., 1999; Halloway et al., 2005) and binds S100B (Figure 1).

A co-immunoprecipitation strategy followed by Western blot analysis demonstrated heterodimerization of S100B and TLR4 in peri-infarct left ventricular myocardium. Interestingly, the RAGE ligand high-mobility group box-1 (HMGB1) a nuclear factor released by necrotic cardiac myocyes following ischemia-reperfusion injury of the heart, also signals via TLR4 (Andassy et al, 2008). Activation of TLR4 results in the activation of NF-κB, which induces the production of proinflammatory cytokines and angiogenic factors thereby promoting inflammation (Beijnum et al., 2007). A total of 10 TLRs have been identified in humans, playing a critical role in both innate and adaptive immunity (Akira et al., 2006) by recognizing and binding to a variety of pathogenic agonists such as lipopeptide (TLR2), double-stranded RNA (TLR3), lipopolysacharide (TLR4), flagellin (TLR5) and deoxycytidylate-phosphate-deoxyguanylate DNA (TLR9) by molecular pattern recognition resulting in the production of inflammatory cytokines in host immune cells (Akira et al., 2006). TLRs are also capable of responding to stress and modulating inflammation and tissue damage following noninfectious insults such as hypoxia and ischemia (Chao et al., 2009). The heart expresses at least six receptors (TLR2, TLR3, TLR4, TLR5, TLR7 and TLR9) (Boyd et al., 2006). The most studied TLR in the heart is TLR4 (Frantz et al., 1999; Knuefermann et al., 2002; Nemoto et al., 2002). Population-based studies designed to determine the impact of TLR4 polymorphism on the risk of myocardial infarction are inconclusive. Some studies suggest that individuals with a single polymorphism of TLR4 have a lower risk of myocardial infarction (Balistreri et al., 2004; Boekholdt et al., 2003; Holloway et al., 2005), whereas others show an increased (Edfeldt et al., 2004) or the same level of risk (Koch et al., 2006). In animal models of ischemic cardiac injury, the role of TLR4 is not well defined. In an *in vivo* model of ischemia-reperfusion injury, mice deficient of TLR4 exhibit reduced myocardial inflammation compared with wild-type mice, suggesting TLR4 mediates ischemic injury in the heart (Chong et al., 2004; Piper et al., 2003). In contrast, the administration of sublethal doses LPS which signals through the TLR4, reduced the subsequent ischemic myocardial infarction and improved cardiac function *in vivo* and in isolated hearts (Brown et al., 1989; Lipton et al., 2001). TLR4 has been linked to both proapototic and survival pathways. The TLR4 ligand LPS induces apoptosis in endothelial cells (Choi et al., 1998) but has an antiapoptotic effect in cardiac myocytes (Zhu et al., 2006). In cells protected by LPS treatment, the activation of TLR4 results in coupling to myeloid differentiation primary-response gene 88 (MyD88) and tumor necrosis factor receptor-associated factor (TRAF) leading to the expression of cell survival and inflammatory genes via NF-κB- and activator protein (AP)-1-dependent mechanisms (Akira and Takeda 2004). The signaling cascades of RAGE and TLR4 show much overlap with the activation of NF-κB. It is tempting to speculate that S100B-TLR4 signaling may trigger a survival pathway in cardiac myocytes to counterbalance the apoptotic pathway activated by S100B-RAGE signaling. The ultimate fate of the cell is dependent on crosstalk with downstream mechanisms leading to activation/repression of signaling pathways.

Glucocorticoid (GCs) Receptors

Crosstalk between GCs and TLR Signaling

The GC receptor is one such downstream target that represses a large set of functionally related inflammatory response genes by disrupting p65/interferon regulatory factor (IRF) complexes required for TLR4-dependent transcriptional activation (Ogawa et al 2005). This mechanism enables the GC receptor to differentially regulate specific programs of gene expression. The GC receptor is a prototypic of a subset of ligand-dependent nuclear receptors that integrate stress and host immune responses with physiological circuits that are required for maintenance of necessary organ functions.

The ability of the GC receptor to repress transcriptional responses to inflammatory signals is an essential component of its homeostatic functions and a primary mechanism by which natural and synthetic GC receptor agonists exert anti-inflammatory effects in a variety of disease settings (Reichart et al 2001).

Negative regulation is thought to result, at least in part, from the ability of GC receptor to interfere by transrepression, with other signal-dependent transcription factors that include NF-κB and activator protein-1 (AP-1) family members (De Bosscher et al 2003). Numerous models have been proposed for GR-mediated transrepression, including direct interactions with NF-κB components (Liden et al 1997), regulation of components of signal transduction pathways involved in NF-κB and AP-1 activation (Caelles et al, 1997), competition for essential co-activators (Sheppard et al, 1998), alternate utilization of co-activators (Rogatsky et al 2001), recruitment of co-repressors (Nissen and Yamamoto 2000) and modification of core transcription factors (Nissen and Yamamoto 2000). It is not known whether there is crosstalk between RAGE signaling and GCs.

GCs – Function

GCs are essential for life and play a vital role in regulating numerous physiological processes including cardiovascular function and metabolism (Chrousos and Kino, 2009). High levels of GCs due either to increased secretion of endogenous steroid or to chronic exogenous treatment have been associated with increased cardiovascular risk (Hadoke et al., 2006, 2009). Moreover, increased GC levels result in visceral obesity, steroid-induced diabetes and hypercholesterolemia (Rhen and Cidlowski, 2005), which are known cardiovascular risk factors. Of note, endogenous or exogenous Cushing syndrome (known to be associated with hypercortisolaemia) is linked with increased cardiovascular morbidity and mortality (Friedman et al., 1996; Chrousos and Kino, 2009).

Cumulative chronic or intermittent stress may lead to high cortisol levels, development of insulin resistance, dyslipidemia, chronic inflammation, arterial hypertension, and/or diabetes mellitus type 2. These disturbancies promote inflammatory processes in endothelial cells and the development of atherosclerosis, and cardiovascular disease (Chrousos, 2004; Chrousos and Kino, 2009).

In contrast, numerous studies investigating the role of GCs in atherosclerosis (Hadoke et al., 2006, 2009) have shown that glucocorticoids interact with cells of the cardiovascular system and modulate the inflammatory response to injury, while glucocorticoid administration has resulted in reduction of atherogenesis and restenosis in animal models (Hadoke et al., 2009).

Glucocorticoid Receptor Gene and Protein Structure

Glucocorticoids (with cortisol being the major GC in humans) mediate their effects via binding to two types of intracellular nuclear receptors, the glucocorticoid receptor (GR) (low affinity, type 2 corticosteroid receptors) and the mineralocorticoid receptor (MR) (high affinity, type 1 corticosteroid receptors) (Chrousos and Kino, 2009). GRs belong to the nuclear receptor 3 subfamily (Nuclear Receptors Nomenclature Committee, 1999) and regulate gene expression (Van der Laan and Meijer, 2008; Zanchi et al., 2010). The human GR gene (NR3C1) is located on chromosome 5q31–q32 (Hollenberg et al., 1985) and contains 8 translated exons (2–9) and 9 untranslated alternative first exons (Turner and Muller 2005; Turner et al., 2010). Although a single gene encodes human GR (Yudt and Cidlowski, 2002), accumulating evidence has shown the presence of several variants, as a result of alternative transcript splicing and alternative translation initiation (Lu and Cidlowski, 2004; Barnes, 2006). Briefly, there are two GR isoforms, the classical GRα and the GRβ, both generated by the same gene consisting of eight common exons and two different ninth exons (9α and 9β respectively) by alternative splicing near the end of primary transcript (Figure 2). The resulting GR proteins have identical the first 727 amino acids with GRα containing additional 50 amino acids (GRα being a 777 amino acid protein) and GRβ containing additional nonhomologous 15 amino acids (GRβ being a 742 amino acid protein) (Yudt and Cidlowski, 2001; Lu and Cidlowski, 2006; Oakley and Cidlowski, 2010). Additionally, alternative splicing at other GR regions results in splice variants such as GRγ mRNA, GR-A and GR-P. Little is known currently about the tissue expression and function of the aforementioned splice variants (Lu and Cidlowski, 2006; Oakley and Cidlowski, 2010). Alternative translation initiation sites give rise to multiple GRα isoforms (GRα-A, GRα-B, GRα-C1, GRα-C2, GRα-C3, GRα-D1, GRα-D2, and GRα-D3) with truncated amino-termini (Lu and Cidlowski, 2005; Oakley and Cidlowski, 2010). The GRα-A is the classic, full length 777-amino acid human glucocorticoid receptor that is generated from the first initiator codon. These GRα isoforms have similar affinity for GCs and interact similarly with GREs upon ligand activation (Lu et al., 2007; Oakley and Cidlowski, 2010). In general, GR protein structure contains three distinct functional regions: a N-terminal domain (NTD), a central DNA-binding domain (DBD), and a ligand-binding domain (LBD) (Figure 2) (Gross et al., 2009). The N-terminal region contains a major *trans*-activation domain, also called activation function-1 (AF-1) (Van der Laan and Meijer, 2008). The DBD consists of 60–70 amino acids (Gross et al., 2009) and is responsible for binding to the glucocorticoid response elements (GREs) present in genes specifically regulated by GCs. The C-terminal region (~250 amino acids), also called activation function-2 (AF-2), is responsible for hormone binding. Finally, a region between the central DBD and the C-terminal domain is important for the translocation of the activated receptor to the nucleus following glucocorticoid binding (reviewed in Zanchi et al., 2010).

Mechanism of GR Action

Glucocorticoids signal through genomic and non-genomic pathways. The classic, genomic actions of glucocorticoids are mediated through cytosolic GR (Gross and Cidlowski, 2008). Glucocorticoids diffuse readily across cell membranes and bind to GRs in the cytoplasm (Barnes 2006). Cytoplasmic GRs are normally bound to molecular chaperones, such as heat shock protein (HSP) 90 and immunophilins, such as FK506-binding protein 52 (FKBP52), which protect the receptor and prevent its nuclear localisation by covering the sites on the receptor that are needed for transport across the nuclear membrane into the nucleus (Pratt and Toft, 1997; Toft, 1998; Wu et al., 2004; Barnes 2006). Upon binding of glucocorticoid, GR undergoes conformational changes, dissociates from the HSPs, homodimerizes and translocates to the nucleus, where it binds to GREs onto DNA in the promoter of target genes, and act as transcription factor by regulating the transcriptional activity of specific target genes in a cell type-specific manner (Schaaf and Cidlowski, 2003). Thus, within minutes of glucocorticoid binding to its receptor, the GR as homodimer enters the nucleus and activates or represses the transcription of target genes (Picard and Yamamoto, 1987; Diamond et al., 1990; Yang-Yen et al., 1990; Yudt and Cidlowski, 2002; Freedman and Yamamoto, 2004). Moreover, the GR as monomer may interact with other transcription factors such as nuclear factor kappa B (NF-κβ), activator protein-1 (AP-1), cyclic AMP response element binding protein (CREB), signal transducer and activator of transcription 5 (STAT-5), and prevent the binding of these transcription factors to their target genes sequences, resulting thus in repression of gene transcription (Dostert and Heinzel, 2004; Zanchi et al., 2010). This mechanism is assumed to mediate anti-inflammatory and immune-suppressive effects of glucocorticoids (Stahn and Buttgereit, 2008; Zanchi et al., 2010). Co-regulatory proteins such as the co-activators p160, p300/CREB binding protein (CPB) and p300/CPB-associated factor (P/CAF), and the co-repressors are important modulators of GR induced gene expression (Kumar and Thompson, 2003). Co-activators and co-repressors are enzymatically active proteins which modulate chromatin structure facilitating transcription initiation or transcription termination (Zanchi et al., 2010). Co-activator proteins usually have histone acetyl transferase (HAT) activity, which facilitates transcription initiation, while co-repressor proteins have histone deacetylase (HDAC) activity, which facilitates transcription termination (Métivier et al., 2006; Zanchi et al., 2010).

GRα is by far the most active form of the receptor, it binds to glucocorticoids, whereas GRβ binds to DNA but cannot be activated by GCs and is thought to be a dominant negative regulator of the GRα (Oakley and Cidlowski, 2010). Recent data, however, on global gene expression analyses have shown that GRβ can directly induce and repress a large number of genes not controlled by GRα (Kino et al., 2009a, 2009b; Lewis-Tuffin et al., 2007; Oakley and Cidlowski, 2010), while the GRβ-induced repression of target genes is attributed to its ability to constitutively induce histone deacetylation (Kelly et al., 2008; Kim et al., 2009; Oakley and Cidlowski, 2010).

Post-Translational Modification of GR Isoforms

Post-translational modification of GR, such as phosphorylation, acetylation, ubiquitination and sumoylation governs greatly GR subcellular localization, the receptor stability, protein expression level and function. GRα is phosphorylated by various kinases, such as the mitogen activated protein kinases (MAPKs), cyclin dependent kinases (CDKs), and glycogen synthase kinase-3 (GSK-3) at the N-terminal domain of the receptor (Oakley and Cidlowski, 2010) resulting in impaired or enhanced GR-induced gene expression. The phosphatases PP1, PP2a, and PP5 reverse the GR phosphorylation regulating thus GR signaling cascades and tissue responsiveness to GCs. Ubiquitination of GRα at a conserved lysine residue in NTD leads the receptor to the proteasome for degradation (Deroo et al., 2002; Wallace and Cidlowski, 2001; Oakley and Cidlowski, 2010). Sumoylation of GRα, at specific lysine residues (Lys-277, Lys-293, and Lys-703) (Oakley and Cidlowski, 2010), decreases the GR-induced gene expression via recruitment of corepressors (Davies et al., 2008; Holmstrom et al., 2003, 2008; Iñiguez-Lluhí and Pearce, 2000; Le Drean et al., 2002; Lin et al., 2006; Tian et al., 2002; Oakley and Cidlowski, 2010), while promotes the receptor degradation. Acetylation of GR at Lys-494 and Lys-495 greatly affects the GR-NFκB interaction (Ito et al., 2006). Recent evidence indicates that miRNAs are involved in the regulation of hippocampal GR levels modulating the HPA axis responsiveness (de Kloet et al., 2009; Turner et al., 2010). The 3′untranslated region (UTR) of the GR was considered to contain numerous seed regions recognised by a variety of miRNAs two of which were miR-18 and miR-124a (Uchida et al., 2008; Turner et al.,

2010). In animal studies, Vreugdenhil et al (2009) found that miRs 18 and 124a reduced GR-mediated events in addition to decreasing GR protein levels. Moreover, they report that: miR-18 and -124a overexpression reduced GR protein levels; miR-18 and -124a overexpression attenuated GR-mediated transactivation; miR-18 and -124a overexpression reduced the induction of the well-known GR-target gene GILZ; miR-124a is able to bind to the predicted seed region in the GR 3′UTR (Vreugdenhil et al., 2009). The miR-124a is recognised as one of 32 ''CNS-heart-shared''-specific miRNAs (Tang et al., 2007). 'Prereceptor cortisol metabolism' by 11β-hydroxysteroid dehydrogenases (HSDs) (11β-HSD1 and 11β-HSD2) controls the intracellular bioavailability of cortisol (Gross and Cidlowski, 2008). 11β-hydroxysteroid dehydrogenase type 2 (11β-HSD2) catalyzes the conversion of cortisol to the inert 11-keto metabolite cortisone, the inactive glucocorticoid metabolite, whereas 11β-hydroxysteroid dehydrogenase type 1 (11β-HSD1) regenerating active cortisol from inert cortisone and thereby amplifying GR activation (Seckl and Walker, 2001; Tomlinson et al., 2004; Morton and Seckl, 2008; Gross and Cidlowski, 2008). 11β-HSD1 and 11β-HSD2 have described in vascular smooth muscle and in the endothelium respectively (Walker et al., 1991; Christy et al., 2003; Walker, 2007). Low levels of 11β-HSD1 and 11β-HSD2 have been shown in human and rat heart (Walker et al., 1991; Smith and Krozowski, 1996; Kayes-Wandover and White, 2000; Slight et al., 1996; Bonvalet et al., 1995; Lombes et al., 1995) while 11β-HSD1 is expressed in differentiated macrophages (Thieringer et al., 2001; Gilmour et al., 2006). Studies using non-selective enzyme inhibitors (Teelucksingh et al., 1990; Walker et al., 1992, 1994; Brem et al., 1997), antisense knockdown (Souness et al., 2002) and studies on global 'knockout' mice (Christy et al., 2003; Hadoke et al., 2001) addressing the role of 11β-HSDs on vascular function have concluded that the loss of 11β-HSD1 reductase regeneration of glucocorticoid in vascular smooth muscle did not affect vascular tone, while loss of 11β-HSD2 dehydrogenase inactivation of glucocorticoid in the endothelium was linked with enhanced vasoconstrictor responses (Walker, 2007). Results from recent studies propose that 11β-HSD1 affects remodelling responses in the vasculature (Hermanowski-Vosatka et al., 2005; Small et al., 2005; Walker, 2007).

Non-Genomic Glucocorticoid Actions

Rapid, non-genomic glucocorticoid actions occurring within seconds have also been reported. These rapid GC effects are considered to be mediated through physiochemical interactions with cellular membranes, the cytosolic GR or a membrane-bound GR (intact or variant) or even a possible G-protein coupled receptor (Song and Buttgereit, 2006; Lowenberg et al., 2008). Briefly, GCs intercalation in plasma and mitochondrial membranes rapidly impaires cation transport across the plasma membrane and increases the mitochondrial proton leak. This non-genomic pathway is considered to play a role in glucocorticoid-induced immunosuppression (Song and Buttgereit, 2006). The previously described cytosolic GR might also play a part in the induction of non-genomic glucocorticoid actions while many rapid glucocorticoid actions are thought to occur through membrane-bound GR (intact or variant), however the underlying mechanisms remain largely unknown (Gross and Cidlowski, 2008). In a recent study, membrane-bound GR was identified as a component of the T-cell receptor (TCR) multi-protein complex. GCs rapidly induced dissociation of this complex and disrupted TCR signaling, thus providing additional insight into the non-genomic mechanism of glucocorticoid-induced immunosuppression, specifically in T cells (Lowenberg et al., 2006, 2008). The non-genomic mechanisms of glucocorticoid action warrant future studies because they might emerge as important pharmacological targets (Gross and Cidlowski, 2008).

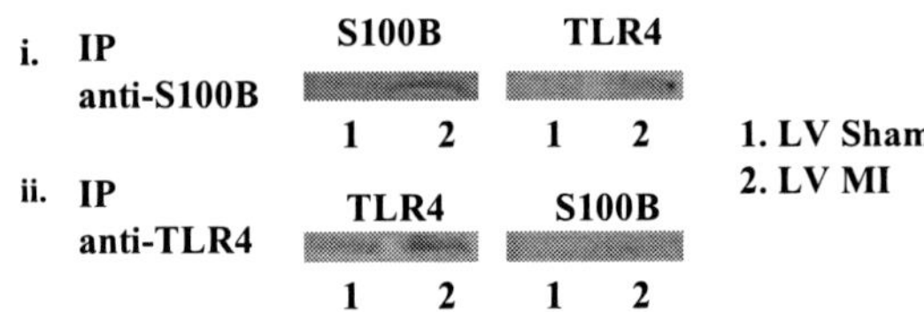

Figure 1. Co-immunoprecipitation of S100B and TLR4 in peri-infarct left ventricular (LV) myocardium 14 days after myocardial infarction. Left anterior descending coronary artery ligation or sham operation was performed on 8 week old male Sprague Dawley rats. Fourteen days after myocardial infarction, tissue lysates from LV myocardium (sham) and peri-infarcted LV myocardium were incubated with either anti-S100B (i) or anti-TLR4 (ii) antibodies in a co-immunoprecipitation assay. The immune complexes were dissociated and analyzed by Western blotting with anti-S100B or anti-RAGE antibodies as indicated above each representative blot.

GR Polymorphisms

Increasing evidence implicates a role of GR gene polymorphisms in cardiovascular disease (Figure 1; detailed in Table 1) (van Rossum and Lamberts, 2004). The ER22/23EK polymorphism is present in exon 2 and results in an arginine (R) to lysine (K) change at position 23 (R23K) within the N terminus (Derijk and de Kloet, 2008). van Rossum and coworkers (2002) in a population-based cohort study in the elderly, found that the ER22/23EK carriers were relatively more resistant to the effects of GCs than noncarriers, resulting in a better metabolic health profile. A recent study has shown that adult carriers of the ER22/23EK polymorphism have a lower risk to develop impaired glucose tolerance, type-2 diabetes and cardiovascular disease. The presence of the ER22/23EK polymorphism in elderly men resulted in lower levels of C-reactive protein, which has been implicated in cardiovascular disease (CAD) (van Rossum et al., 2004a; van Rossum and Lamberts, 2004). In contrast, in a large study with 2024 patients with familial hypercholesterolemia who are known to have a high risk of cardiovascular disease, no significant association of the ER22/23EK variant with cardiovascular disease risk was found, however, a sex-specific risk modification was identified; in males appearing a relative protection, while in females there was a slightly increased risk of cardiovascular disease (Koeijvoets et al., 2006). The N363S polymorphism is also present in the GR N terminus (exon 2) and occurs in ~4% of individuals (Derijk and de Kloet, 2008). Some studies report that N363S is associated with an increased body mass index (BMI) (Huizenga et al., 1998; Lin et al., 1999; Di Blasio et al., 2003) and coronary artery disease (Lin et al., 2003b), while a previous study found an increased waist to hip ratio (WHR) but no association with BMI in N363S male carriers (Dobson et al., 2001). No association was found with hypertension and the N363S polymorphism (Lin et al., 2003a). The *Bcl*I variant is a restriction fragment length polymorphism that occurs in ~37% of individuals and is located within intron 2 (Derijk and de Kloet, 2008). Data, regarding the effect of *Bcl*I on metabolic parameters are controversial (van Rossum and Lamberts, 2004). For instance, associations were found with several cardiovascular risk factors, in carriers of *Bcl*I polymorphism (Rosmond et al., 2000a; Rosmond and Holm, 2008), while Kuningas et al (2006) were found no associations with HDL-and LDL-cholesterol levels, triglycerides, or HbA1c in *Bcl*I carriers. In addition, Ukkola et al (2001), in a study conducted with 12 pairs of monozygotic twins, were described an atherogenic profile in response to overfeeding, in subjects who are homozygotes for the 2.3 kb allele. However, individuals harboring both the *Bcl*I and N363S polymorphism tend to have higher blood pressure and cholesterol levels (Di Blasio et al., 2003). In addition, Watt et al (1992) found an association of *Bcl*I polymorphism with hypertension. Furthermore, Alevizaki et al (2007) have shown that the *Bcl*I variant was linked with increased severity of coronary artery disease concomitantly with high HPA-axis reactivity to stress. High-risk patients with severe familial hypercholesterolemia were investigated for *Bcl*I and 9β haplotypes; in men, but not in women, the *Bcl*I and 9β haplotypes were significantly associated with increased CVD risk (Koeijvoets et al., 2008).

Finally, heterozygotes for another polymorphism within intron 2, rs2918419 (T→C), were identified in ~25.8% of subjects in a population-based study in northeast England (Syed et al., 2008). Men carrying the dual polymorphisms (i.e. rs2918419 and *BclI*) showed insulin resistance (Syed et al., 2008), however, this was not observed in women. The association of *BclI* polymorphism with insulin resistance and obesity seems to be crucially depend on the presence of rs2918419 polymorphism (Gross and Cidlowski, 2008) since in this study in the absence of rs2918419 (Syed et al., 2008) the previously mentioned detrimental effect of the *BclI* polymorphism was not observed. Polymorphism of the glucocorticoid receptor gene (in the 5' end of the GR gene) was found to be associated with elevated cortisol secretion (Björntorp and Rosmond, 2000) and be responsible for the associated insulin resistance, central obesity and hypertension (Björntorp et al., 1999). A *TthIIIΙ* polymorphism (found within the GR promoter, 3807 base pairs upstream of the transcriptional start site) was linked with healthy metabolic profile in the presence of ER22/23EK (van Rossum and Lamberts, 2004; van Rossum et al., 2004). In a previous study, Rosmond et al (2000b) was found an association of *TthIIIΙ* polymorphism with elevated diurnal cortisol levels. A GR gene intron 4 polymorphism G1666T, which found in Chinese individuals, contributes to the development of cerebral infarction (CI) in females and G allele may be a predisposing gene marker (Chi et al., 2003). The GRβ polymorphism A3669G (located within the 3′ untranslated region) results in increased stability of *GRβ* mRNA and the enhanced expression of GRβ protein (Gross and Cidlowski, 2008). The A3669G allele has been linked with reduced central obesity in women and a more favorable lipid profile in men (Syed et al., 2006), implicating that some of the undesirable effects of GRα on fat distribution and lipid metabolism may be antagonized by a rise in GRβ (Oakley and Cidlowski, 2010). In contrast, the beneficial immunosuppressive and anti-inflammatory actions of GRα, attenuated by GRβ elevation may also underlie the increased risk of A3669G carriers for pathologies with known inflammatory components such as autoimmune disease, myocardial infarction, and coronary artery disease (van den Akker et al., 2008). A recent study on European population showed that women heterozygous for A3669G had decreased incidence of central obesity (Syed et al., 2006), while men heterozygous for A3669G had decreased total cholesterol concomitant with elevated high-density lipoprotein. Since other similar studies and the A3669G carriers from South Asia did not demonstrate similar findings, it is suggested that A3669G might not be a reliable predictor of metabolic profiles across all ethnic groups (Syed et al., 2006; van den Akker et al., 2006). Recently, A3669G homozygous carriers were associated with a pro-inflammatory phenotype that included an increased risk of myocardial infarction and coronary heart disease (CHD) (van den Akker et al., 2008). More recently, Otte et al (2010) report that in a cohort of 526 A3669G carriers with stable CHD the 9β polymorphism was associated with prevalent HF, systolic dysfunction, and subsequent HF hospitalization in patients with CHD and was partly mediated by low-grade inflammation.

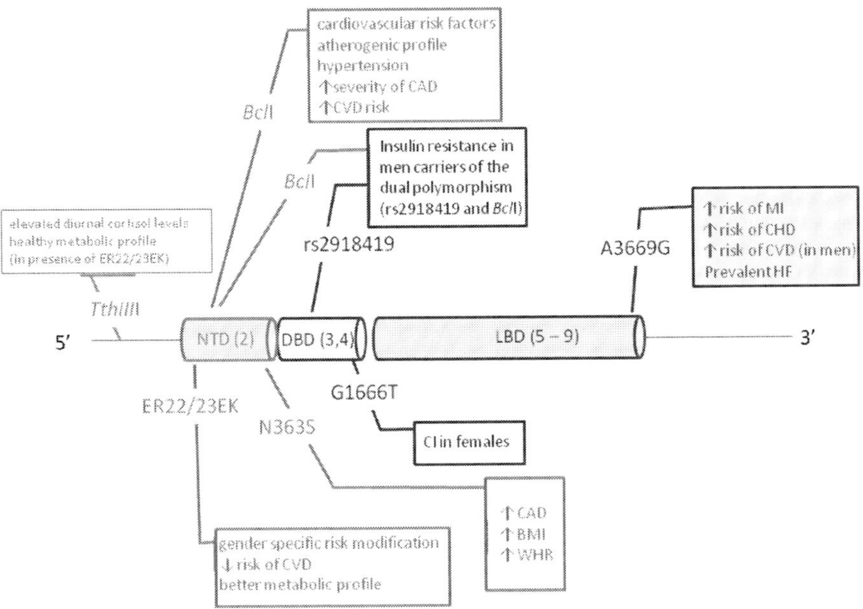

Figure 2. Human GR primary transcript. Common polymorphisms [*TthIII*I, ER22/23EK, N363S, *Bcl*I, rs2918419, G1666T, A3669G (5' to 3')] in the human glucocorticoid receptor gene and their associations with cardiovascular disease and cardiovascular risk factors. Inset: Transactivation, DNA binding, ligand binding; the hGR primary transcript contains 9 exons, with exon 2 encoding most of the amino-terminal transactivation (NTD) domain, exons 3-4 encoding the DNA-binding domain (DBD), and exons 5-9 encoding the hinge region and ligand-binding domain (LBD). [Modified from Gross and Cidlowski, 2008].

Concluding Remarks

In conclusion, one can propose a model whereby S100 proteins, in particular S100B is induced in response to inflammatory signals (e.g. ischemia-reperfusion injury) and in addition to its intracellular actions, is released into the extracellular space by damaged cardiac myocytes. S100B via ligand binding to either/both TLR4 and RAGE trigger signaling pathways resulting in the activation of the transcription factors NF-κB/AP-1 leading to the stimulation of GC receptors resulting in transrepression of signal dependent target genes (Figure 3).

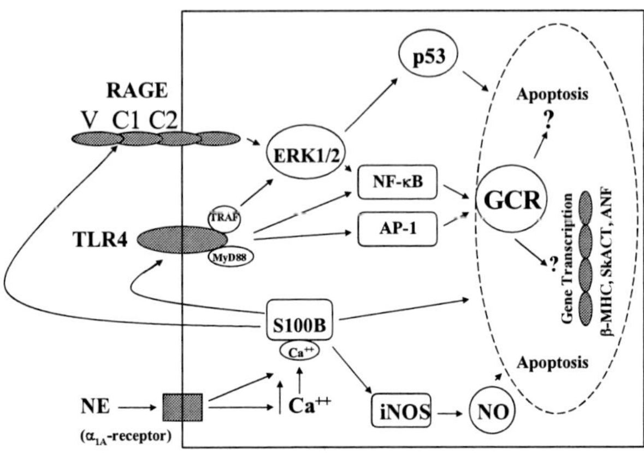

Figure 3. Schematic of S100B signaling in cardiac myocytes. Norepineprine (NE) via α_{1A} receptor binding induces intracellular S100B. S100B, activated by intracellular calcium (Ca^{++}) binds components of the protein kinase C (PKC)-β signaling pathway resulting in inhibition of gene [(e.g. β-myosin heavy chain (MHC), skeletal α actin (skACT), atrial natriuretic factor (ANF)] transcription. Intracellular S100B can induce apoptosis via an inducible nitric oxide synthase (iNOS)-NO pathway. Secreted S100B via activation of the receptor for advanced glycation end products (RAGE) (extracellular components V and CI) can also induce apoptosis via MEK-ERK1/2-p53 signaling. RAGE signaling can activate nuclear factor kappa-B (NF-κB) and potentially the glucocorticoid receptors (GCR) leading to a transrepression of target genes. S100B can also activate toll-receptor (TLR)-4 and via signaling through myeloid differentiation primary-response gene 88 (MyD88) and tumor necrosis factor receptor-associated factor (TRAF) can activate NF-κB and activator protein (AP)-1 and potentially the glucocorticoid receptors (GCR) again leading to a transrepression of target genes. The common signaling pathway involves the activation of NF-κB which in turn can activate GCRs.

Abbreviations

CVD, cardiovascular disease; CAD, coronary artery disease; HF, heart failure; CHD, coronary heart disease; MI, myocardial infarction; cerebral infarction (CI); FH, familial hypercholesterolemia; BMI, body mass index; WHR, waist-to-hip ratio.

REFERENCES

Alevizaki M, Cimponeriu A, Lekakis J, Papamichael C, Chrousos GP. High anticipatory stress plasma cortisol levels and sensitivity to glucocorticoids predict severity of coronary artery disease in subjects undergoing coronary angiography. *Metabolism,* (2007) 56(2):222-226.

Akira S, Takeda K. Toll-like receptor signalling. *Nat. Rev. Immunol.* (2004) 4:499-511.

Akira S, Uematsu S, Takeuchi O. Pathogen recognition and innate immunity. *Cell,* (2006) 124:783-801.

Andassy M, Volz HC, Igwe JC, Funkle B, Eichberger SN, KayaZ, Buss S, Autschbach F, Pleger ST, Lukic IK, Bea F, Hardt SE, Humpert PM, Bianchi ME, Mairbaural H, Nawroth PP, Rempiss A, Katus HA, Bierhaus A. High-mobility group box-1 in ischemia-reperfusion injury of the heart. *Circulation,* (2008) 117:3216-3226.

Anderson RE, Hansson LO, Vaage J. Release of S100B during coronary artery bypass grafting is reduced by off-pump surgery. *Ann. Thorac. Surg.* (1999) 67:1721-1725.

Balistreri CR, Candore G, Colonna-Romano G, Lio D, Caruso M, Hoffman E, Franceschi C, Caruso C. Role of Toll-like receptor 4 in acute myocardial infarction and longevity. *JAMA* (2004) 292:2339-2340.

Barnes PJ. Corticosteroid effects on cell signalling. *Eur. Respir. J.* (2006) 27:413–426.

Baudier J, Cole RD. Interactions between the microtubule-associated tau proteins and S100b regulate tau phosphorylation by the Ca2+/calmodulin-dependent protein kinase II. *J. Biol. Chem.* (1988) 263:5876-5883.

Becker T, Gerke V, Kube E, Weber K. S100P, a novel Ca(2+)-binding protein from human placenta. cDNA cloning, recombinant protein expression and Ca2+ binding properties. *Eur. J. Biochem.* (1992) 207:541-547.

Bierhaus A, Humpert PM, Morcos T, Wendt T, Chavakis T, Arnold B, Stern DM, Nawroth PP. Understanding RAGE, the receptor for advanced glycation end products. *J. Mol. Med.* (2005) 83:876-886.

Björntorp P, Holm G, Rosmond R. Hypothalamic arousal, insulin resistance and Type 2 diabetes mellitus. *Diabet. Med.* (1999) 16(5):373-383.

Bjorntorp P, Rosmond R. Obesity and cortisol. *Nutrition,* (2000) 16:924–936.

Boekholdt SM, Agema WR, Peters RJ, Zwinderman AH, van der Wall EE, Reitsma PH, Kastelein JJ, Jukema JW. Variants of toll-like receptor 4 modify the efficacy of statin therapy and the risk of cardiovascular events. *Circulation,* (2003) 107:2416-2421.

Bonvalet JP, Alfaidy N, Farman N, Lombes M. Aldosterone: intracellular receptors in human heart. *Eur. Heart J.* (1995) 16:92–97.

Boyd JH, Mathur S, Wang Y, Bateman RM, Walley KR. Toll-like receptor stimulation in cardiomyocytes decreases contractility and initiates an NF-kappaB dependent inflammatory response. *Cardiovasc. Res.* (2006) 72:384-393.

Brem AS, Bina RB, Hill N, Alia C, Morris DJ. Effects of licorice derivatives on vascular smooth muscle function. *Life Sci.* (1997) 60:207–214.

Brown JM, Grosso MA, Terada LS, Whitman GJ, Banerjee A, White CW, Harken AH, Repine JE. Endotoxin pre-treatment increases endogenous myocardial catalase activity and decreases ischemia-reperfusion injury of isolated rat hearts. *Proc. Natl. Acad. Sci. USA,* (1989) 86:2516-2520.

Caelles C, Ganzalez-Sancho JM, Munoz A. Nuclear hormone receptor antagonism with AP-1 by inhibition of the JNK pathway. *Genes Dev.* (1997) 11:3351-3364.

Chao W. Toll-like receptor signalling; a critical modulator of cell survival and ischemic injury in the heart. *Am. J. Physiol. Heart Circ. Physiol.* (2009) 296:H1-H12.

Chi LQ, Zhang C, Luo B, Sun HY. The association between glucocorticoid receptor gene G1666T polymorphism and cerebral infarction in Chinese. *Zhonghua Yi Xue Yi Chuan Xue Za Zhi* (2003) 20(4):353-356. [Article in Chinese].

Choi KB, Wong F, Harlaan JM, Chaudhary PM, Hood L, Karsan A. Lipopolysaccharide mediated endothelial apoptosis by a FADD-dependent pathway. *J. Biol. Chem.* (1998) 273:20185-20188.

Chong AJ, Shimamoto A, Hampton CR, Takayama H, Spring DJ, Rothnie CL, Yada M, Pohlman TH, Verrier ED. Toll-like receptor 4 mediates ischemia /reperfusion injury of the heart. *J. Thorac. Cardiovasc. Surg.* (2004) 128:170-179.

Christy C, Hadoke PWF, Paterson JM, Mullins JJ, Seckl JR, Walker BR. Glucocorticoid action in mouse aorta; localisation of 11b-hydroxysteroid dehydrogenase type 2 and effects on responses to glucocorticoids in vitro. *Hypertension* (2003) 42:580–587.

Chrousos GP. The glucocorticoid receptor gene, longevity, and the complex disorders of Western societies. *Am. J. Med.* (2004) 117:204–207.

Chrousos GP, Kino T. Glucocorticoid Signaling in the Cell: Expanding Clinical Implications to Complex Human Behavioral and Somatic Disorders. *Ann. NY Acad. Sci.* (2009) 1179: 153–166.

Davies L, Karthikeyan N, Lynch JT, Sial EA, Gkourtsa A, Demonacos C, Krstic-Demonacos M. Cross talk of signaling pathways in the regulation of the glucocorticoid receptor function. *Mol. Endocrinol.* (2008) 22(6):1331-1344.

De Bosscher K, Vanden Berghe W, Haegeman G. The interplay between the gluccocorticoid and nuclear factor-κB or activator protein-1:molecular mechanisms for gene repression. *Endocr. Rev.* (2003) 24:488-452.

de Kloet ER, Fitzsimons CP, Datson NA, Meijer OC, Vreugdenhil E. Glucocorticoid signaling and stress-related limbic susceptibility pathway: about receptors, transcription machinery and microRNA. *Brain Res.* (2009) 1293:129–141.

Delphin C, Ronjat M, Deloulme JC, Garin G, Debussche L, Higashimoto Y, Sakaguchi K, Baudier J. Calcium-dependent interaction of S100B with the C-terminal domain of the tumor suppressor p53. *J. Biol. Chem.* (1999) 274:10539-10544.

Derijk RH, de Kloet ER. Corticosteroid receptor polymorphisms: determinants of vulnerability and resilience. *Eur. J. Pharmacol.* (2008) 583:303–311.

Deroo BJ, Rentsch C, Sampath S, Young J, DeFranco DB, Archer TK. Proteasomal inhibition enhances glucocorticoid receptor transactivation and alters its subnuclear trafficking. *Mol. Cell Biol.* (2002) 22 (12):4113-4123.

Diamond MI, Miner JN, Yoshinaga SK, Yamamoto KR. Transcription factor interactions: selectors of positive or negative regulation from a single DNA element. *Science,* (1990) 249:1266–1272.

Di Blasio AM, van Rossum EF, Maestrini S, Berselli ME, Tagliaferri M, Podestà F, Koper JW, Liuzzi A, Lamberts SW. The relation between two polymorphisms in the glucocorticoid receptor gene and body mass index, blood pressure and cholesterol in obese patients. *Clin. Endocrinol.* (Oxf) (2003) 59:68–74.

Donato R. Functional roles of S100 proteins, calcium-binding proteins of the EF-hand type. *Biochim. Biophys. Acta,* (1999) 1450:191-231.

Donato R. S100: a multigenic family of calcium-modulated proteins of the EF-hand type with intracellular and extracellular functional roles. *Int. J. Biochem. Cell Biol.* (2001) 33:637-668.

Donato R. Intracellular and extracellular roles of S100 proteins. *Microsc. Res. Tech.* (2003) 60:540-551.

Donato R. RAGE: a single receptor for several ligands and different cellular responses: the case of certain S100 proteins. *Curr. Mol. Med.* (2007) 7:711-724.

Donato R, Sorci G, Riuzzi F, Arcuri C, Bianchi R, Brozzi F, Tubaro C, Giambanco I (2009) S100B's double life: Intracellular regulator and extracellular signal. Intracellular and extracellular roles of S100 proteins. *Biochim. Biophys. Acta,* (2009) 1793:1008-1022.

Dobson MG, Redfern CP, Unwin N, Weaver JU. The N363S polymorphism of the glucocorticoid receptor: potential contribution to central obesity in men and lack of association with other risk factors for coronary heart disease and diabetes mellitus. *J. Clin. Endocrinol. Metab.* (2001) 86(5):2270-2274.

Dostert A, Heinzel T. Negative glucocorticoid receptor response elements and their role in glucocorticoid action. *Curr. Pharm. Des.* (2004) 10:2807–2816.

Edfeldt K, Bennet AM, Eriksonn P, Frostegard J, Wiman B, Hamsten A, Hansson GK, de Faire U, Yan ZQ. Association of hypo-responsive toll-like receptor 4 variants with risk of myocardial infarction *Eur. Heart J.* (2004) 25:1447-1453.

Emoto Y, Kobayashi R, Akatsuka H, Hidaka H. Purification and characterization of a new member of the S-100 protein family from human placenta. *Biochem. Biophys. Res. Commun.* (1992) 182:1246-1253.

Engelkamp D, Schafer BW, Mattei MG, Erne P, Heizmann CW. Six S100 genes are clustered on human chromosome 1q21: identification of two genes coding for the two previously unreported calcium-binding proteins S100D and S100E. *Proc. Natl. Acad. Sci. USA,* (1993) 90:6547-6551.

Frantz S, Kobzik L, Kim YD, Fukazawa R, Medzhitov R, Lee RT, Kelly RA. Toll4 (TLR4) expression in cardiac myocytes in normal and failing myocardium. *J. Clin. Invest.* (1999) 104:271-280.

Freedman ND, Yamamoto KR. Importin 7 and importin alpha/importin beta are nuclear import receptors for the glucocorticoid receptor. *Mol. Biol. Cell* (2004) 15:2276–2286.

Friedman TC, Mastorakos G, Newman TD, Mullen NM, Horton EG, Costello R, Papadopoulos NM, Chrousos GP. Carbohydrate and lipid metabolism in endogenous hypercortisolism: shared features with metabolic syndrome X and NIDDM. *Endocr. J.* (1996) 43:645–655.

Frizzo JK, Tramontina F, Bortoli E, Gottfried C, Leal RB, Lengyel I, Donato R, Dunkley PR, Goncalves CA. S100B-mediated inhibition of the phosphorylation of GFAP is prevented by TRTK-12. *Neurochem. Res.* (2004) 29:735-740.

Gilmour JS, Coutinho AE, Cailhier JF, Man TY, Clay M, Thomas G, Harris HJ, Mullins JJ, Seckl JR, Savill JS, Chapman KE. Local amplification of glucocorticoids by 11beta-hydroxysteroid dehydrogenase type 1 promotes macrophage phagocytosis of apoptotic leukocytes. *J. Immunol.* (2006) 176:7605–7611.

Goncalves DS, Lenz G, Karl J, Goncalves CA, Rodnight R. Extracellular S100B protein modulates ERK in astrocyte cultures. *Neuroreport,* (2000) 11:807-809.

Griffin WS, Sheng JG, McKenzie JE, Royston MC, Gentleman SM, Brumback RA, Cork LC, Del Bigio MR, Roberts GW, Mrak RE. Life-long overexpression of S100beta in Down's syndrome: implications for Alzheimer pathogenesis. *Neurobiol. Aging,* (1998) 19:401-405.

Gross KL, Cidlowski JA. Tissue-specific glucocorticoid action: a family affair *Trends Endocrinol. Metab.* (2008) 19(9): 331–339.

Gross KL, Lu NZ, Cidlowski JA. Molecular mechanisms regulating glucocorticoid sensitivity and resistance. *Mol. Cell Endocrinol.* (2009) 300:7–16.

Hadoke PWF, Macdonald L, Logie JJ, Small GR, Dover AR, Walker BR. Intravascular glucocorticoid metabolism as a modulator of vascular structure and function. *Cell Mol. Life Sci.* (2006) 63:565–578.

Hadoke PWF, Iqbal J, Walker BR. Therapeutic manipulation of glucocorticoid metabolism in cardiovascular disease. *Bri. J. Pharmacol.* (2009) 156:689–712.

Heizmann CW, Ackermann GE, Galichet A. Pathologies involving the S100 proteins and RAGE. *Subcel. Biochem.* (2007) 45:93-108.

Heizmann CW and Cox JA. New perspectives on S100 proteins: a multi-functional Ca(2+)-, Zn(2+)- and Cu(2+)-binding protein family. *Biometals,* (1998) 11:383-397.

Hermanowski-Vosatka A, Balkovec JM, Cheng K, Chen HY, Hernandez M, Koo GC, Le Grand CB, Li Z, Metzger JM, Mundt SS, Noonan H, Nunes CN, Olson SH, Pikounis B, Ren N, Robertson N, Schaeffer JM, Shah K, Springer MS, Strack AM, Stowski M, Wu K, Wu T, Xiao J, Zhang BB, Wright SD, Thieringer R. 11β-HSD1 inhibition ameliorates metabolic syndrome and prevents progression of atherosclerosis in mice. *J. Exp. Med.* (2005) 202 517–527.

Hollenberg SM, Weinberger C, Ong ES, Cerelli G, Oro A, Lebo R, Thompson EB, Rosenfeld MG, Evans RM. Primary structure and expression of a functional human glucocorticoid receptor cDNA. *Nature,* (1985) 318(6047):635-641.

Holloway JW, Yang IA, Ye S. Variation in the toll-like receptor 4 gene and susceptibility to myocardial infarction. *Pharmacogenet. Genomics,* (2005) 15:15-21.

Holmstrom S, Van Antwerp ME, Iniguez-Lluhi JA. Direct and distinguishable inhibitory roles for SUMO isoforms in the control of transcriptional synergy. *Proc. Natl. Acad. Sci. USA* (2003) 100(26):15758-15763.

Holmstrom SR, Chupreta S, So AY, Iniguez-Lluhi JA. SUMO-mediated inhibition of glucocorticoid receptor synergistic activity depends on stable assembly at the promoter but not on DAXX. *Mol. Endocrinol.* (2008) 22(9):2061-2075.

Hu J, Van Eldik LJ. S100 beta induces apoptotic cell death in cultured astrocytes via a nitric oxide - dependent pathway. *Biochim. Biophys. Acta,* (1996) 1313:239-245.

Huizenga NA, Koper JW, De Langc P, Pols HA, Stolk RP, Burger H, Grobbee DE, Brinkmann AO, De Jong FH, Lamberts SW. A polymorphism in the glucocorticoid receptor gene may be associated with and increased sensitivity to glucocorticoids *in vivo. J. Clin. Endocrinol. Metab.* (1998) 83:144–151.

Huttunen HJ, Kuja-Panula J, Sorci G, Agneletti AL, Donato R, Rauvala H. Coregulation of neurite outgrowth and cell survival by amphoterin and S100 proteins through receptor for advanced glycation end products (RAGE) activation. *J. Biol. Chem.* (2000) 275:40096-40105.

Ikura M. Calcium binding and conformational response in EF-hand proteins. *Trends Biochem. Sci.* (1996) 21:14-17.

Iñiguez-Lluhí JA, Pearce D. A common motif within the negative regulatory regions of multiple factors inhibits their transcriptional synergy. *Mol. Cell Biol.* (2000) 20(16): 6040-6050.

Ito K, Yamamura S, Essilfie-Quaye S, Cosio B, Ito M, Barnes PJ, Adcock IM. Histone deacetylase 2-mediated deacetylation of the glucocorticoid receptor enables NF-kappaB suppression. *J. Exp. Med.* (2006) 203(1):7-13.

Iuvone T, Esposito G, De Filippis D, Bisogna T, Petrosino S, Scuderi C, DiMarzo V, Steardo L. Cannabinoid CB1 receptor stimulation affords neuroprotection in MPTP-induced neurotoxicity by attenuating S100B up-regulation in vitro. *L. Mol. Med.* (2007) 85:1379-1392.

Karin M, Ben-Neriah Y. Phosphorylation meets ubiquitination: the control of NF-[kappa B] activity. *Annu. Rev. Immunol.* (2000) 18:621-663.

Kayes-Wandover KM, White PC. Steroidogenic enzyme gene expression in the human heart. *J. Clin. Endocrinol. Metab.* (2000) 85:2519–2525.

Kelly A, Bowen H, Jee YK, Mahfiche N, Soh C, Lee T, Hawrylowicz C, Lavender P. The glucocorticoid receptor beta isoform can mediate transcriptional repression by recruiting histone deacetylases. *J. Allergy Clin. Immunol.* (2008) 121(1):203-208 e201.

Kerkhoff C, Klempt M, Sorg C. Novel insights into structure and function of MRP8 (S100A8) and MRP14 (S100A9). *Biochim. Biophys. Acta,* (1998)1448:200-211.

Kilby PM, Van Eldik LJ, Roberts GC. Identification of the binding site on S100B protein for the actin capping protein CapZ. *Protein Sci.* (1997) 6:2494-2503.

Kim SH, Kim DH, Lavender P, Seo JH, Kim YS, Park JS, Kwak SJ, Jee YK. Repression of TNF-alpha-induced IL-8 expression by the glucocorticoid receptor-beta involves inhibition of histone H4 acetylation. *Exp. Mol. Med.* (2009) 41(5):297-306.

Kino T, Su Y A, Chrousos GP. Human glucocorticoid receptor isoform beta: recent understanding of its potential implications in physiology and pathophysiology. *Cell Mol. Life Sci.* (2009a) 66(21):3435-3448.

Kino T, Manoli I, Kelkar S, Wang Y, Su YA, Chrousos GP. Glucocorticoid receptor (GR) beta has intrinsic, GRalpha-independent transcriptional activity. *Biochem. Biophys. Res. Commun.* (2009b) 381(4):671-675.

Kligman D and Marshak DR. Purification and characterization of a neurite extension factor from bovine brain. *Proc. Natl. Acad. Sci. USA,* (1985) 82:7136-7139.

Knuefermann P, Nemoto S, Misra A, Nozaki N, Defreitas G, Goyert SM, Carabello BA, Mann DL, Vallejo JG. CD14-deficient mice are protected against lipopolysaccharide-induced cardiac inflammation and left ventricular dysfunction. *Circulation,* (2004) 110:1693-3698.

Koch W, Hoppmann P, PfeuferA, Schomig A, Kastrati A. Toll-like receptor 4 polymorphisms and myocardial infarction; no association in a Caucasian population. *Eur. Heart J.* (2006) 27:2524-2529.

Koeijvoets KC, van Rossum EF, Dallinga-Thie GM, Steyerberg EW, Defesche JC, Kastelein JJ, Lamberts SW, Sijbrands EJ. A functional polymorphism in the glucocorticoid receptor gene and its relation to cardiovascular disease risk in familial hypercholesterolemia. *J. Clin. Endocrinol. Metab.* (2006) 91: 4131–4136.

Koeijvoets KCMC, van der Net JB, van Rossum EFC, Steyerberg EW, Defesche JC, Kastelein JJP, Lamberts SWJ, Sijbrands EJG. Two Common Haplotypes of the Glucocorticoid Receptor Gene Are Associated with Increased Susceptibility to Cardiovascular Disease in Men with Familial Hypercholesterolemia. *J. Clin. Endocrinol. Metab.* (2008) 93:4902–4908.

Kumar R, Thompson EB. Transactivation functions of the N-terminal domains of nuclear hormone receptors: protein folding and coactivator interactions. *Mol. Endocrinol.* (2003) 17:1–10.

Kursula P, Lehto VP, Heape AM. S100beta inhibits the phosphorylation of the L-MAG cytoplasmic domain by PKA. *Mol. Brain Res.* (2000) 76:407-410.

Kuningas M, Mooijaart SP, Slagboom EP, Westendorp RGJ, van Heemst D Genetic variants in the glucocorticoid receptor gene (NR3C1) and cardiovascular disease risk. The Leiden 85-plus Study. *Biogerontology,* (2006) 7:231–238.

Leclerc E, Fritz G, Vetter SW, Heizmann CW. Binding of S100 proteins to RAGE: an update. *Biochim. Biophys. Acta,* (2009) 1793:993-1007.

Leclerc E, Fritz G, Weibel M, Heizmann CW, Galichet A. S100B and S100A6 differentially modulate cell survival by interacting with distinct RAGE immunoglobulin domains. *J. Biol. Chem.* (2007) 282:31317-31331.

Le Drean Y, Mincheneau N, Le Goff P, Michel D. Potentiation of glucocorticoid receptor transcriptional activity by sumoylation. *Endocrinology,* (2002) 143(9):3482-3489.

Liden J, Franck D, Rafter I, Guftafson JA, Okret S. A new function for the C-terminal zinc finger of the glucocorticoid receptor; regression of RelA transactivation. *J. Biol. Chem.* (1997) 272:21467-21472.

Lin DY, Huang YS, Jeng JC, Kuo HY, Chang CC, Chao TT, Ho CC, Chen YC, Lin TP, Fang HI, Hung CC, Suen CS, Hwang MJ, Chang KS, Maul GG, Shih HM. Role of SUMO-interacting motif in Daxx SUMO modification, subnuclear localization, and repression of sumoylated transcription factors. *Mol. Cell,* (2006) 24(3):341-354.

Lin RC, Wang WY, Morris BJ. High penetrance, overweight, and glucocorticoid receptor variant: case-control study. *BMJ,* (1999) 319:1337–1338.

Lin RC, Wang XL, Morris BJ. Association of coronary artery disease with glucocorticoid receptor N363S variant. *Hypertension* (2003b) 41:404–407.

Lin RC, Wang XL, Dalziel B, Caterson ID, Morris BJ. Association of obesity, but not diabetes or hypertension, with glucocorticoid receptor N363S variant. *Obes. Res.* (2003a) 11:802–808.

Lipton BP, Delcarpio JB, McDonough KH. Effects of endotoxin on neutriphil-mediated I/R injury in isolated perfused rat hearts. *Exp. Biol. Med.* (2001) 226:320-327.

Lewis-Tuffin LJ, Jewell CM, Bienstock RJ, Collins JB, Cidlowski JA. Human glucocorticoid receptor beta binds RU-486 and is transcriptionally active. *Mol. Cell Biol.* (2007) 27(6):2266-2282.

Lombes M, Alfaidy N, Eugene E, Lessana A, Farman N, Bonvalet J-P. Prerequisite for cardiac aldosterone action: mineralocorticoid receptor and 11beta-hydroxysteroid dehydrogenase in the human heart. *Circulation,* (1995) 92:175–182.

Löwenberg M, Verhaar AP, Bilderbeek J, Marle J, Buttgereit F, Peppelenbosch MP, van Deventer SJ, Hommes DW. Glucocorticoids cause rapid dissociation of a T-cell-receptor-associated protein complex containing LCK and FYN. *EMBO Rep.* (2006) 7:1023–1029.

Löwenberg M, Stahn C, Hommes DW, Buttgereit F. Novel insights into mechanisms of glucocorticoid action and the development of new glucocorticoid receptor ligands. *Steroids,* (2008) 73:1025–1029.

Lu NZ, Cidlowski JA. The origin and functions of multiple human glucocorticoid receptor isoforms. *Ann. N.Y. Acad. Sci.* (2004) 1024:102–123.

Lu NZ, Cidlowski JA. Translational regulatory mechanisms generate N-terminal glucocorticoid receptor isoforms with unique transcriptional target genes. *Mol. Cell,* (2005) 18(3):331-342.

Lu NZ, Cidlowski JA. Glucocorticoid receptor isoforms generate transcription specificity. *Trends Cell Biol.* (2006) 16:301–307.

Lu NZ, Collins JB, Grissom SF, Cidlowski JA. Selective regulation of bone cell apoptosis by translational isoforms of the glucocorticoid receptor. *Mol. Cell Biol.* (2007) 27(20):7143-7160.

Markowitz J, Rustandi RR, Varney KM, Wilder PT, Udan R, Wu SL, Horrocks WD, Weber DJ. Calcium-binding properties of wild-type and EF-hand mutants of S100B in the presence and absence of a peptide derived from the C-terminal negative regulatory domain of p53. *Biochemistry,* (2005) 44:7305-7314.

Mazzini GS, Schaf DV, Vinade ER, Horowitz E, Bruch RS, Brunm LM, Goncalves CA, Bacal F, Souza DO, Portela LV, Bordignon S. Increased S100B serum levels in dilated cardiomyopathy patients. *J. Card. Fail.* (2007) 13:850-854.

Mbele GO, Deloulme JC, Gentil BJ, Delphin C, Ferro M, Garin J, Takahashi M, Baudier J. The zinc- and calcium-binding S100B interacts and co-localizes with IQGAP1 during dynamic rearrangement of cell membranes. *J. Biol. Chem.* (2002) 277: 49998-50007.

Millward TA, Heizmann CW, Schafer BW, Hemmings BA.Calcium regulation of Ndr protein kinase mediated by S100 calcium-binding proteins. *EMBO J.* (1998)17:5913-5922.

Métivier R, Reid G, Gannon F. Transcription in four dimensions: Nuclear receptor directed initiation of gene expression. *EMBO J.* (2006) 7:161–167.

Morton NM, Seckl JR. 11beta-hydroxysteroid dehydrogenase type 1 and obesity. *Front Horm. Res.* (2008) 36:146–164.

Nemoto s, Vallejo JG, Knuefermann P, Misra A, Defrietas G, Carbello BA, Mann DL. Escheria coli LPS-induced LV dysfunction:role of toll-like receptor-4 in the adult heart. *Am. J. Physiol. Heart Circ. Physiol.* (2002) 382:H2316-H2323.

Newton RA and Hogg N. The human S100 protein MRP-14 is a novel activator of the beta 2 integrin Mac-1 on neutrophils. *J. Immunol.* (1998) 160:1427-1435.

Nissen RM, Yamamoto KR. The glucocorticoid receptor inhibits NF-κB by interfering with serine-2phosphorylation of the RNA polymerase II carboxy-terminal-domain. *Genes Dev.* (2000) 14:2314-2329.

Oakley RH, Cidlowski JA. Cellular Processing of the Glucocorticoid Receptor Gene and Protein: New Mechanisms for Generating Tissue Specific Actions of Glucocorticoids. *J. Biol. Chem.* (2011) 286:3177-3184.

Ogawa S, Lozach J, Benner C, Pascual G, Tangirala RK, Westin S, Hoffmann A, Subramaniam S, David M, Rosenfeld MG, Glass CK. *Cell,* (2005) 122:707-721.

Ostendorp T, Leclerc E, Galichet A, Koch M, Demling N, Weigle B, Heizmann CW, Kroneck PM. Structural and functional insights into RAGE activation by multimeric S100B. *EMBO J.* (2007) 26:3868-3878.

Otte C, Wüst S, Zhao S, Pawlikowska L, Kwok PY, Whooley MA. Glucocorticoid receptor gene, low-grade inflammation, and heart failure: the Heart and Soul study. *J. Clin. Endocrinol. Metab.* (2010) 95(6):2885-2891.

Peskind ER, Griffin WS, Akama KT, Raskind MA,Van Eldik LJ. Cerebrospinal fluid S100B is elevated in the earlier stages of Alzheimer's disease. *Neurochem. Int.* (2001) 39:409-413.

Picard D, Yamamoto KR. Two signals mediate hormone-dependent nuclear localization of the glucocorticoid receptor. *EMBO J.* (1987) 6:3333–3340.

Piper HM, Meuter K, Schafer C. Cellular mechanisms of ischemia- reperfusion injury. *Ann. Thorac. Surg.* (2003) 75:s644-s648.

Portela LV, Brenol JC, Walz R, Bianchin M, Tort AB, Canabarro UP, Beheregaray S, Marasca JA, Xavier RM, Neto EC, Goncalves CA, Souza DO. Serum S100B levels in patients with lupus erythematosus: preliminary observation. *Clin. Diagn. Lab. Immunol.* (2002) 9:164-166.

Pratt WB, Toft DO. Steroid receptor interactions with heat shock protein and immunophilin chaperones. *Endocr. Rev.* (1997) 18:306–360.

Razavi HM, Hamilton JA, Feng Q. Modulation of apoptosis by nitric oxide: implications in myocardial ischemia and heart failure. *Pharmacol. Rev.* (2005) 106:147-162.

Reichardt HM, Tuckermann JP, Gottlicher M, Vujic M, Weih F, Angel P, Herrlich P, Schultz G. Repression of inflammatory responses in the absence of DNA binding by the glucocorticoid receptor. *EMBO J.* (2001) 20:7168-7173.

Rhen T, Cidlowski JA. Antiinflammatory action of glucocorticoids — new mechanisms for old drugs. *N. Engl. J. Med.* (2005) 353:1711–1723.

Rogatsky I, Zarember KA, Yamamoto KR. Factor recruitment and TIF2/GRIP1 corepressor activity at a collagenase-3 response element that mediates regulation by phorbol esters and hormones. *EMBO J.* (2001) 20:7168-7173.

Rosmond R, Chagnon YC, Holm G, Chagnon M, Pérusse L, Lindell K, Carlsson B, Bouchard C, Björntorp P. A glucocorticoid receptor gene marker is associated with abdominal obesity, leptin, and dysregulation of the hypothalamic-pituitary-adrenal axis. *Obes. Res.* (2000a) 8:211-218.

Rosmond R, Chagnon YC, Chagnon M, Pérusse L, Bouchard C, Björntorp P. A polymorphism of the 5'-flanking region of the glucocorticoid receptor gene locus is associated with basal cortisol secretion in men. *Metabolism,* (2000b) 49(9):1197-1199.

Rosmond R, Holm G. A 5-year follow-up study of 3 polymorphisms in the human glucocorticoid receptor gene in relation to obesity, hypertension, and diabetes. *J. Cardiometab. Syndr.* (2008) 3(3):132-5.

Rustandi RR, Baldisseri DM, Weber DJ. Structure of the negative regulatory domain of p53 bound to S100Bb. *Nat. Struct. Biol.* (2000) 7:570-574.

Santamaria-Kisiel L, Rintala-Dempsey AC, Shaw GS. Calcium-dependent and -independent interactions of the S100 protein family. *Biochem. J.* (2006) 396:201-214.

Schaaf MJ, Cidlowski JA. Molecular determinants of glucocorticoid receptor mobility in living cells: the importance of ligand affinity. *Mol. Cell Biol.* (2003) 23:1922–1934.

Schaub MC, Heizmann CW. Calcium, troponin, calmodulin, S100 proteins: from myocardial basics to new therapeutic strategies. *Biochem. Biophys. Res. Commun.* (2008) 369:247-264.

Schmidt AM, Yan SD, Yan SF, Stern DM. The multiligand receptor RAGE as a progression factor amplifying immune and inflammatory responses. *J. Clin. Invest.* (2001) 108:949-955.

Seckl JR, Walker BR. 11β-hydroxysteroid dehydrogenase type 1– a tissue-specific amplifier of glucocorticoid action. *Endocrinology*, (2001) 142:1371–1376.

Slight SH, Ganjam VK, Gomez-Sanchez CE, Zhou M-Y, Weber KT. High affinity NADC-dependent 11beta-hydroxysteroid dehydrogenase in the human heart. *J. Mol. Cell. Cardiol.* (1996) 28:781–787.

Small GR, Hadoke PWF, Sharif I, Dover AR, Armour D, Kenyon CJ, Gray GA, Walker BR. Preventing regeneration of glucocorticoids by 11bhydroxysteroid dehydrogenase type 1 enhances angiogenesis. *Proc. Natl. Acad. Sci. USA,* (2005) 102:12165–12170.

Smith RE, Krozowski ZS. 11β-hydroxysteroid dehydrogenase type 1 enzyme in the hearts of normotensive and spontaneously hypertensive rats. *Clin. Exp.Pharmacol. Physio.* (1996) 23:642–647.

Song IH, Buttgereit F. Non-genomic glucocorticoid effects to provide the basis for new drug developments. *Mol. Cell Endocrinol.* (2006) 246:142–146.

Souness GW, Brem AS, Morris DJ. 11beta-hydroxysteroid dehydrogenase antisense affects vascular contractile response and glucocorticoid metabolism. *Steroids*, (2002) 67:195–201.

Stahn C, Buttgereit F. Genomic and nongenomic effects of glucocorticoids. *Nat. Clin. Pract. Rheumatol.* (2008) 4:525–533.

Syed AA, Irving JA, Redfern CP, Hall AG, Unwin NC, White M, Bhopal RS, Weaver JU. Association of glucocorticoid receptor polymorphism A3669G in exon 9β with reduced central adiposity in women. *Obesity,* (2006)14:759–764.

Syed AA, Halpin CG, Irving JA, Unwin NC, White M, Bhopal RS, Redfern CP, Weaver JU. A common intron 2 polymorphism of the glucocorticoid receptor gene is associated with insulin resistance in men. *Clin. Endocrinol.* (2008) 68(6):879–884.

Tang X, Gal J, Zhuang X, Wang W, Zhu H, Tang G. A simple array platform for microRNA analysis and its application in mouse tissues. *RNA* (2007) 13:1803–1822.

Teelucksingh S, Mackie ADR, Burt D, McIntyre MA, Brett L, Edwards CRW. Potentiation of hydrocortisone activity in skin by glycyrrhetinic acid. *Lancet,* (1990) 335:1060–1063.

Thieringer R, Le Grand CB, Carbin L, Cai T-Q, Wong B, Wright SD. 11β-hydroxysteroid dehydrogenase type 1 is induced in human monocytes upon differentiation to macrophages. *J. Immunol.* (2001) 167:30–35.

Tian S, Poukka H, Palvimo JJ, Janne OA. Small ubiquitin-related modifier-1 (SUMO-1) modification of the glucocorticoid receptor. *Biochem. J.* (2002) 367(Pt 3):907-911.

Toft DO. Recent advances in the study of hsp90 structure and mechanism of action. *Trends Endocrinol. Metab.* (1998) 9:238–243.

Tomlinson JW, Walker EA, Bujalska IJ, Draper N, Lavery GG, Cooper MS, Hewison M, Stewart PM. 11b-Hydroxysteroid dehydrogenase type 1: a tissue-specific regulator of glucocorticoid response. *Endocr. Rev.* (2004) 25:831–866.

Tsoporis JN, Marks A, Kahn HJ, Butany JW, Liu PP, O'Hanlon D, Parker TG. S100B inhibits α_1-adrenergic induction of the hypertrophic phenotype in cardiac myocytes. *J. Biol. Chem.* (1997) 272:31915-31921.

Tsoporis JN, Marks A, Kahn HJ, Butany JW, Liu PP, O'Hanlon D, Parker TG. Inhibition of norepinephrine-induced cardiac hypertrophy in S100beta transgenic mice. *J. Clin. Invest.* (1998) 102:1609-1616.

Tsoporis JN, Marks A, Haddad A, Dawood F, Liu PP, Parker TG. S100B expression modulates left ventricular remodeling after myocardial infarction in mice. *Circulation,* (2005) 111:598-606.

Tsoporis JN, Overgaard CB, Izhar S, Parker TG. S100B modulates the hemodynamic response to norepinephrine stimulation. *Am. J. Hypertens.* (2009) 22:1048-1053.

Tsoporis JN, Izhar S, Leong-Poi H, Desjardins J-F, Huttunen HJ, Parker TG. S100B Interaction with the receptor for advanced glycation end products (RAGE). A novel receptor-mediated mechanism for myocyte apoptosis postinfarction. *Circ. Res.* (2010) 106:93-101.

Turner JD, Muller CP. Structure of the glucocorticoid receptor (NR3C1) gene 50 untranslated region: identification, and tissue distribution of multiple new human exon 1. *J. Mol. Endocrinol.* (2005) 35:283–292.

Turner JD, Alt SR, Cao L, Vernocchi S, Trifonova S, Battello N, Muller CP. Transcriptional control of the glucocorticoid receptor: CpG islands, epigenetics and more. *Biochem. Pharmacol.* (2010) 80:1860–1868.

Uchida S, Nishida A, Hara K, Kamemoto T, Suetsugi M, Fujimoto M, Watanuki T, Wakabayashi Y, Otsuki K, McEwen BS, Watanabe Y. Characterization of the vulnerability to repeated stress in Fischer 344 rats: possible involvement of microRNA-mediated down-regulation of the glucocorticoid receptor. *Eur. J. Neurosci.* (2008) 27(9):2250–2261.

Ukkola O, Rosmond R, Tremblay A, Bouchard C. Glucocorticoid receptor *Bcl*I variant is associated with an increased atherogenic profile in response to long-term overfeeding. *Atherosclerosis,* (2001) 157(1):221-224.

van Beijnum JR, Buurman WA. Convergence and amplification of toll-like receptor (TLR) and receptor for advanced glycation end products (RAGE) signaling pathways via high mobility group B1 (HMGB1). *Angiogenesis,* (2008) 11:91-99.

van den Akker EL, Russcher H, van Rossum EF, Brinkmann AO, de Jong FH, Hokken A, Pols HA, Koper JW, Lamberts SW. Glucocorticoid receptor polymorphism affects transrepression but not transactivation. *J. Clin. Endocrinol. Metab.* (2006) 91(7):2800–2803.

van den Akker EL, Koper JW, van Rossum EF, Dekker MJ, Russcher H, de Jong FH, Uitterlinden AG, Hofman A, Pols HA, Witteman JC, Lamberts SW. Glucocorticoid receptor gene and risk of cardiovascular disease. *Arch. Intern. Med.* (2008) 168(1):33-39.

van der Laan S, Meijer OC. Pharmacology of glucocorticoids: Beyond receptors. *Eur. J. Pharmacol.* (2008) 585:483–491.

van Eldik LJ and Griffin WS (1994) S100 beta expression in Alzheimer's disease: relation to neuropathology in brain regions. *Biochim. Biophys. Acta,* (1994) 1223:398-403.

van Eldik LJ and Zimmer DB. Secretion of S-100 from rat C6 glioma cells. *Brain Res.* (1987) 436: 367-370.

van Rossum EFC, Koper JW, Huizenga NATM, Uitterlinden AG, Janssen JAMJL, Brinkmann AO, Grobbee DE, de Jong FH, van Duyn CM, Pols HAP, Lamberts SWJ. A polymorphism in the glucocorticoid receptor gene, which decreases sensitivity to glucocorticoids in vivo, is associated with low insulin and cholesterol levels. *Diabetes,* (2002) 51:3128–3134.

van Rossum EF, Lamberts SW. Polymorphisms in the glucocorticoid receptor gene and their associations with metabolic parameters and body composition. *Recent Prog. Horm. Res.* (2004) 59:333–357.

van Rossum EF, Feelders RA, van den Beld AW, Uitterlinden AG, Janssen JA, Ester W, Brinkmann AO, Grobbee DE, de Jong FH, Pols HA, Koper JW, Lamberts SW. Association of the ER22/23EK polymorphism in the glucocorticoid receptor gene with survival and C-reactive protein levels in elderly men. *Am. J. Med.* (2004) 117:158–162.

Vreugdenhil E, Verissimo CSL, Mariman R, Kamphorst JT, Barbosa JS, Zweers T, Champagne DL, Schouten T, Meijer OC, Ron de Kloet E, Fitzsimons CP. MicroRNA 18 and 124a Down-Regulate the Glucocorticoid Receptor: Implications for Glucocorticoid Responsiveness in the Brain. *Endocrinology,* (2009) 150: 2220–2228.

Walker BR, Yau JL, Brett LP, Seckl JR, Monder C, Williams BC, Edwards CRW. 11β-Hydroxysteroid dehydrogenase in vascular smooth muscle and heart: implications for cardiovascular responses to glucocorticoids. *Endocrinology,* (1991) 129:3305–3312.

Walker BR, Connacher AA, Webb DJ, Edwards CRW. Glucocorticoids and blood pressure: a role for the cortisol/cortisone shuttle in the control of vascular tone in man. *Clin. Sci.* (1992) 83:171–178.

Walker BR, Sang KS, Williams BC, Edwards CRW. Direct and indirect effects of carbenoxolone on responses to glucocorticoids and noradrenaline in rat aorta. *J. Hypertens.* (1994) 12:33–39.

Walker BR. Glucocorticoids and cardiovascular disease. *Eur. J. Endocrinol.* (2007) 157:545–559.

Wallace AD, Cidlowski JA. Proteasome-mediated glucocorticoid receptor degradation restricts transcriptional signaling by glucocorticoids. *J. Biol. Chem.* (2001) 276(46):42714-42721.

Watt GC, Harrap SB, Foy CJ, Holton DW, Edwards HV, Davidson HR, Connor JM, Lever AF, Fraser R. Abnormalities of glucocorticoid metabolism and the renin-angiotensin system: a four-corners approach to the identification of genetic determinants of blood pressure. *J. Hypertens.* (1992) 10:473–482.

Wu B, Li P, Liu Y, et al. 3D structure of human FK506-binding protein 52: implications for the assembly of the glucocorticoid receptor/Hsp90/immunophilin heterocomplex. *Proc. Natl. Acad. Sci. USA,* (2004) 101:8348–8353.

Yang-Yen HF, Chambard JC, Sun YL, Smeal T, Schmidt TJ, Drouin J, Karin M. Transcriptional interference between c-Jun and the glucocorticoid receptor: mutual inhibition of DNA binding due to direct protein-protein interaction. *Cell,* (1990) 62:1205–1215.

Yudt MR, Cidlowski JA. Molecular identification and characterization of a and b forms of the glucocorticoid receptor. *Mol. Endocrinol.* (2001) 15(7):1093-1103.

Yudt MR, Cidlowski JA. The glucocorticoid receptor: coding a diversity of proteins and responses through a single gene. *Mol. Endocrinol.* (2002) 16:1719–1726.

Zanchi NE, De Siqueira Filho MA, Felitti V, Nicastro H, Lorenzeti FM, Lancha JR AH. Glucocorticoids: Extensive Physiological Actions Modulated Through Multiple Mechanisms of Gene Regulation. *J. Cell Physiol.* (2010) 224:311–315.

Zhu X, Zhao H, Graveline AR, Buys ES, Schmidt U, Bloch KD, Rosenzweig A, Chao W. MyD88 and NOS2 are essential for toll-like receptor 4-mediated survival effect in cardiomyocytes. *Am. J. Physiol. Heart Circ. Physiol.* (2006) H1900-H1909.

In: Nuclear Receptors
Editors: M. K. Bates and R. M. Kerr
ISBN: 978-1-61209-980-4

Chapter 2

STEROID RECEPTOR COACTIVATORS AND ENDOCRINE TREATMENT IN BREAST CANCER

***Line L. Haugan Moi*[1,3]*, Marianne Hauglid Flågeng*[1,2]*, Simon Nitter Dankel*[1,2]*, Tuyen Hoang*[1,2]*, Jennifer Gjerde*[1,2]*, Jørn V. Sagen*[1,2]*, Ernst A. Lien*[1,2] *and Gunnar Mellgren*[1,2]**

[1]Institute of Medicine, University of Bergen
[2]The Hormone Laboratory, Haukeland University Hospital, N-5021 Bergen
[3]Department of Clinical Pathology, The University Hospital of North Norway, N-9038 Tromsø, Norway

ABSTRACT

Breast cancer is the most frequent malignancy in women in the Western world. Hormone receptor positive breast cancers are managed with endocrine treatment in which the estrogen receptor (ER) is blocked using a selective estrogen receptor modulator (SERM) such as tamoxifen or by targeting the estrogen synthesis using aromatase inhibitors (AIs). Nuclear receptor coactivators have been pointed out as the main determinants of tissue-, cell- and promoter specific effects of tamoxifen, and they are important regulators of ER mediated gene transcription under estrogen deprivation induced by aromatase inhibition.

The steroid receptor coactivator (SRC) family comprises SRC-1, SRC-2/TIF-2 and SRC-3/AIB1. Typically they enhance the transcriptional activity of ligand-bound ER by binding to the activation function-2 (AF-2) pocket, and recruit the basal transcription factors and chromatin-remodeling complex, acetyltransferase proteins, methyltransferases and ubiquitin ligases. 4-hydroxytamoxifen works as an ER antagonist by binding to the nuclear receptor and inducing a displacement of helix 12 that blocks binding of coactivators and favors corepressor recruitment. However, high levels of coactivators relative to corepressors may force ER into an active structural conformation where 4-hydroxytamoxifen leads to ER agonistic effects via AF-1 by indirect binding to DNA and recruitment of coactivators. While SRC-1 and SRC-2/TIF-2 are expressed in normal and malignant breast tissue, SRC-3/AIB1 predominates with overexpression in >30% and gene amplification in 5 – 10% of breast cancers. Our recent studies in human breast cancer have shown that treatment with tamoxifen or AIs enhances gene expression of the SRCs. Others have reported that high levels of SRC-1 are associated with nodal involvement and resistance to endocrine treatment. SRC-3/AIB1 and the growth factor receptor HER-2/neu are often coexpressed in breast cancers, and poor response to tamoxifen treatment and reduced disease-free survival are found when tumors overexpress SRC-1 or SRC-3/AIB1 together with HER-2/neu. The SRCs are regulated by post-translational modifications by for instance mitogen activated protein kinases (MAPKs) which operate downstream of HER-2/neu and stabilize and functionally activate SRC proteins, a mechanism which could contribute not only to tamoxifen resistance, but also to estrogen hypersensitivity and resistance to AIs. In summary, steroid receptor coactivators are crucial in ER regulated gene transcription. Accumulated evidence points to an association between coactivator levels, effect of and response to endocrine treatment and long-term outcome in human breast cancer. The SRCs are involved in crosstalk between ER and growth factor pathways that are activated in breast cancer, making coactivators important in breast cancer development and interesting as potential therapeutic targets.

INTRODUCTION

Breast cancer is the most frequent malignancy and a major cause of cancer deaths in women. The ER belongs to the family of nuclear receptors and is typically considered to be a ligand-regulated and DNA sequence-specific transcription factor which is expressed in >70% of human breast cancers [Osborne et al., 1996]. It is well established that estrogen has pro-carcinogenic effects in mammary epithelium by stimulating proliferation and leaving the cells prone to mutations during cell cycle progression [Foster et al., 2001], in addition to its growth stimulatory effects in breast cancer cells [Lippman and Bolan, 1975]. Hormone receptor positive breast cancers are managed with endocrine treatment. Tamoxifen and its potent metabolites 4-hydroxytamoxifen and 4-hydroxy-*N*-

desmethylamoxifen [Stearns et al., 2003; Lien et al., 1989] bind to and block ER regulated gene transcription whereas AIs inhibit estrogen synthesis. Endocrine treatment decreases mortality and prolongs disease-free survival in women with breast cancer [Early Breast Cancer Trialists' Collaborative Group, 2005], and tamoxifen has been shown to reduce the incidence of ER positive breast cancer in women at increased risk [Cuzick et al., 2003]. The best proof of principle of estradiol as a carcinogen in mammary epithelium is the success of tamoxifen in the prevention and treatment of breast cancer. Nuclear receptor coactivators have been pointed out as the main determinants of tissue- and cell specific effects of tamoxifen. The SRC family includes SRC-1 (also termed nuclear receptor coactivator-1 NCoA), SRC-2/transcription intermediary factor-2 TIF-2 (also termed NCoA-2 and glucocorticoid receptor-interacting protein 1 GRIP1) and SRC-3/amplified in breast cancer 1 AIB1 (also termed NCoA-3, receptor-associated coactivator 3 RAC3, activator of the thyroid and retinoic acid receptor ACTR, thyroid hormone receptor activator molecule-1 TRAM-1 and CBP interacting protein p/CIP). The SRCs are involved in every step of ER regulated transcription, from chromatin remodeling to termination. These coactivators may also modulate signaling through the growth factor receptor HER-2/neu which is overexpressed in 20 – 30 % of breast cancers [Slamon et al., 1987]. The HER-2/neu protein can be targeted in breast cancer therapy using specific antibodies such as trastuzumab or tyrosine kinase inhibitors [Goldhirsch et al., 2009]. It is noteworthy that the SRCs also interact with and coactivate other nuclear receptors as well as other transcription factors such as activation protein-1 (AP-1) [Lee et al., 1998], nuclear factor-kB (NF-kB) [Na et al., 1998] and CREB-binding protein (CBP/p300) [Torchia et al., 1997]. Importantly, coactivators are regulated by post-translational modifications via growth factor signaling pathways that are believed to modulate breast cancer carcinogenesis and endocrine resistance. In this chapter we will focus on the functional role of SRCs in the ERα mediated transcription. Overall, the SRCs are crucial to ER mediated effects and hence to the effects of endocrine treatment in breast cancer. Their expression, regulation and functional roles in breast tissue are therefore of great interest.

Structure, Expression and Function of SRCs in Normal and Malignant Breast Tissue

The SRCs are genetically distinct, but have similar structural and functional properties. They are about 160 kDa with an overall ~40% sequence homology

[Xu and Li, 2003] and similar functional domains (Figure 1). Despite their structural similarities, experimental evidence suggests different physiological functions for the SRCs, which in part can be attributed to tissue-specific expression levels, different preferences for the SRCs between nuclear receptors and post-translational modifications [Hoang et al., 2004; Fenne et al., 2008; Chauchereau et al., 2003; Wu et al., 2004; Xu et al., 2009]. The human SRC-1 gene is located on chromosome 2 (p23) [Carapeti et al., 1998]. The gene is widely expressed in various tissues including brain and nervous system, skeletal and heart muscle, lung, colon, esophagus, conjunctiva and placenta. A notable expression of SRC-1 is seen in various cell types such as B-cells, CD4+ T cells, CD33+ cells, dendritic cells and hematopoietic stem cells [Lukk et al., 2010]. Normal and malignant breast tissue express SRC-1 protein [Hudelist et al., 2003]. The functional role of SRC-1 in ER mediated transcription in breast tissue is indicated by the decreased estrogen-induced mammary gland ductal side branching and alveolar formation in ovariectomized SRC-1$^{-/-}$ mice [Ku et al., 1998]. SRC-1 has also been reported to specifically promote breast cancer metastasis [Wang et al., 2009]. The mRNA and protein levels of SRC-1 were found to be higher in breast cancer compared to normal tissue [Haugan Moi et al., 2010; Hudelist et al., 2003], and SRC-1 protein expression in breast cancer was found to associate positively with HER-2/neu and negatively with ERβ [Fleming et al., 2004].

A smaller study found reduced SRC-1 mRNA levels in recurrent breast cancer [Berns et al., 1998]. Other disease states associated with an altered expression of SRC-1 include B-cell lymphoma, ulcerative colitis, sarcoma, Alzheimer's disease, inflammatory myopathy and placental choriocarcinoma [Lonard et al., 2010; Lukk et al., 2010].

Human SRC-2/TIF-2 is found on chromosome 8 (q21) [Kalkhoven et al., 1998] and is highly expressed in tissues and cell types such as adipose tissue, skeletal muscle, heart, brain, bladder, esophagus, placenta, mediastinum, myometrium, hematopoietic stem cells, T-cells and fibroblasts [Lukk et al., 2010]. An altered expression of SRC-2/TIF-2 is associated with a variety of diseases including hyperparathyroidism, malignant peripheral nerve sheath tumor, sarcoma, heart disease, Alzheimer's disease, head and neck squamous cell carcinoma and polycystic ovarian syndrome [Lonard et al., 2010; Lukk et al., 2010]. The mRNA levels of SRC-2/TIF-2 were found to be elevated in intraductal carcinoma [Kurebayashi et al., 2000] and invasive breast cancer [Haugan Moi et al., 2010] compared to normal breast tissue, but overall SRC-2/TIF-2 is the least studied of the steroid receptor coactivators in breast cancer.

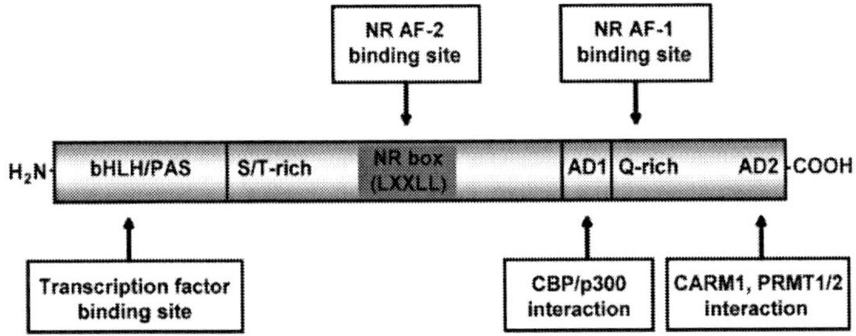

Figure 1. Schematic representation of the structural and functional domains of the steroid receptor coactivators. SRC-1, SRC-2/TIF-2 and SRC-3/AIB1 share a common structure. They have a basic helix-loop-helix/Per-ARNT-Sim homologous domain (bHLH/PAS), a serine/threonine (S/T) rich region and a nuclear receptor (NR) interacting domain with three LXXLL motifs. The short sequence motif LXXLL, where L is leucine and X is any amino acid, is critical for the estrogen/ligand-dependent interaction with ER. The two activation domains AD1 and AD2 are located in the C-terminus where AD1 is responsible for interaction with the general transcriptional machinery through CBP and p300, and recruits acetyltransferases for chromatin remodeling including CBP/p300 and p/CAF. AD2 interacts with the histone methyltransferases coactivator-associated arginine methyltransferase 1 (CARM1) and protein arginine methyltransferase 1 and 2 (PRMT1/2).

In contrast, SRC-3/AIB1 was first identified as a gene amplified and overexpressed in several human breast cancer cell lines and is of great interest in the breast cancer research field [Anzick et al., 1997]. The functional role of SRC-3/AIB1 in normal breast development has been illustrated by knock-out experiments in mice where female SRC-3/AIB1$^{-/-}$ mice have significantly lower levels of estrogen and delayed mammary gland development, indicating a proliferative role of this coactivator in breast tissue [Xu et al., 2000]. In addition to its role in normal mammary development, the *SRC-3/AIB1* gene, located at chromosome 20q, is known to be amplified in 5-10% of breast tumors [Anzick et al., 1997; Bautista et al., 1998] and overexpressed at the mRNA and protein level in 20-60 % of breast cancers, including both ER positive and negative tumors [Anzick et al., 1997; Murphy et al., 2000; Bouras et al., 2001; List et al., 2001]. Overexpression of SRC-3/AIB1 in transgenic mice leads to hypertrophy and hyperplasia in the mammary glandular tissue and development of malignant breast tumors, whereas mice lacking SRC-3/AIB1 are resistant to chemical carcinogen-induced mammary tumorigenesis [Torres-Arzayus et al., 2004; Kuang et al., 2005]. The coactivator is now considered to be a *bona fide* oncogene [reviewed in Lydon and O'Malley, 2010; Gojis et al., 2009; Yan et al., 2006]. High levels of SRC-3/AIB1 mRNA in breast cancer, estimated by in situ hybridization, have been shown to negatively correlate with ER protein levels, but associate with high protein levels of HER-2/neu and the tumor suppressor p53 [Bouras et al., 2001;

Lahusen et al., 2009]. Overall, most studies seem to indicate higher levels of the SRCs in malignant breast tumors compared to normal breast tissue, which is in line with observations made in other endocrine sensitive tissues such as endometrium [Kershah et al., 2004].

SRCs and Activation of ER

The ER is a nuclear receptor that contains the common structural and functional features of a steroid/thyroid hormone receptor (Figure 2). The transcriptional activity of ER is induced when agonists such as estradiol bind to the receptor. It has been well established that binding of agonist to ER induces a conformational change in the receptor's ligand binding domain (LBD), leading to receptor dimerization and association with estrogen response elements (EREs) located on the promoter of target genes. Upon binding of agonist, the helix 12 is positioned across the LBD forming a favorable surface for SRC binding where residues within helix 12 and helices 3 and 5 are essential for SRC-ER interactions [Brzozowski et al., 1997, Mak et al., 1999]. The SRCs typically bind to and enhance ER transcriptional activity in a ligand-dependent manner, by acting as adapters for direct or indirect recruitment of other transcriptional cofactors.

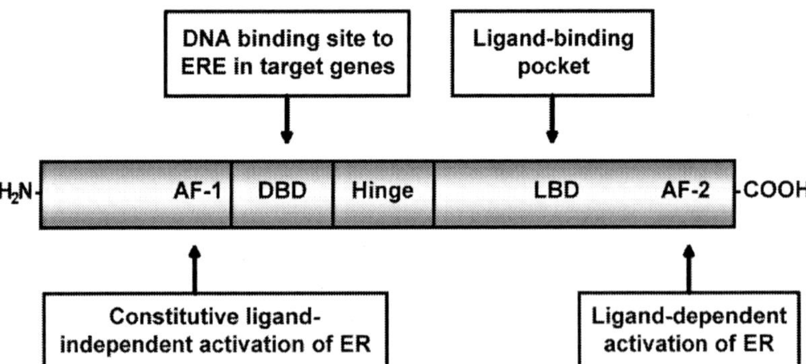

Figure 2. Schematic representation of the structural and functional domains of the estrogen receptor. The autonomous activation function 1 (AF-1) is located in the N-terminus whereas the centrally-located DNA-binding domain (DBD) comprises two zinc finger-like motifs and is responsible for specific binding of ER to estrogen response elements (EREs) on its target genes. The ligand-binding domain (LBD) is composed of 12 helices and harbors a ligand-binding pocket and the activation function 2 (AF-2). The AF-1 of ER contributes to the constitutive estrogen-independent activation by the receptor and is separated by a hinge region from the AF-2, which is structurally conserved and plays a critical role in the estrogen-dependent activation pathway of ER.

These cofactors participate in histone modification, chromatin-remodeling, formation of stable preinitiation complexes and RNA polymerase reinitiation, and include the acetyltransferase proteins CBP/p300/pCAF (p300/CBP-associated factor), methyltransferases CARM1/PRMT1, and ubiquitin ligases E6-AP [Xu and O'Malley, 2002]. ER can also regulate transcription of genes that do not harbor a classical ERE, such as *c-MYC* and *IGF-1,* by indirect binding to DNA via other transcription factors such as SP1 and AP-1 [Dubik et al., 1992; Umayahara et al., 1994].

Estrogen functions as a growth factor in breast cancer by promoting recruitment of coactivators after binding to ER. The coactivators can contribute to proliferation and breast carcinogenesis by contributing to ER-regulated mechanisms, including stimulation of the G1 to S phase transition during cell cycle by regulating the expression of mitogenic genes such as *c-FOS, c-MYC* and *cyclin D1* [Dubik et al., 1992; Weisz et al., 1990; Sabbah et al., 1999]. SRC-dependent activation of ER also stimulates expression of growth factors such as TGF-α and IGF-I, the growth factor receptor EGFR and genes involved in chromatin remodeling, thereby facilitating gene transcription [DiRenzo et al., 2000]. The ER regulated *c-MYC* is an important regulator of proliferation, cell growth and apoptosis. c-MYC mediates resistance to apoptosis and activates cyclin-dependent kinases which promote cell cycle progression, and has been linked to endocrine resistance [McNeil et al., 2006]. It is noteworthy that estrogen can stimulate proliferation, not only by transcriptional regulation, but also through activation of protein kinases and second messengers by membrane bound ER and crosstalk with other growth factor receptor signaling pathways [Kelly and Levin, 2001; Yager and Davidson, 2006].

The oncogenic potential of SRC-3/AIB1 has been ascribed to mechanisms such as increased interaction between ER and the promoter of the ER target gene *cyclin D1*, leading to increased levels of cyclin D1 and stimulation of cell cycle progression [Planas-Silva et al., 2001]. Accordingly, the gene expression of cyclin D1 was found to be reduced in SRC-3/AIB1 knock-out MCF-7 breast cancer cells [Karmakar et al., 2009], and mice with reduced SRC-3/AIB1 expression show a decrease in epithelial proliferation linked to a reduction in cyclin D1 and cyclin E expression [Fereshteh et al., 2008]. Knock-out experiments in MCF-7 cells have demonstrated that RNA depletion of SRC-2/TIF-2 or SRC-3/AIB1, but not of SRC-1, inhibited growth of the cells due to decreased cell cycle progression, increased apoptosis and reduced ER mediated transcriptional activity [Karmakar et al., 2009]. However, the SRCs have been shown to contribute to cancer development through both estrogen dependent- and independent functions [Torres-Arzayus et al., 2010].

SRCs and Cancer-Related Factors

Transgenic mice overexpressing SRC-3/AIB1 develop tumors in several organs including breast, and demonstrate increased expression of IGF-1 and activation of the IGF-IR/PI3K/Akt pathway [Torres-Arzayus et al., 2004]. Overexpression of SRC-3/AIB1 also stimulates the Akt signaling pathway and promotes cell growth [Torres-Arzayus et al., 2004; Zhou et al., 2003]. Amplification of *SRC-3/AIB1* has been found to associate with high levels of MDM2 that negatively regulates the important tumor suppressor p53 which is found defective in many human cancers, including breast cancer [Bautista et al., 1998]. High mRNA levels of SRC-3/AIB1 have also been found to associate with high protein levels of p53, where the high p53 levels may signify an oncogenic event with inactivation of the important tumor suppressor protein p53 [Bouras et al., 2001]. Both SRC-1 and SRC-3/AIB1 have been found to coactivate the MAPK dependent transcription factor ETS-2 which regulates the expression of *c-MYC* [Al-asawi et al., 2008]. In a clinical study on patients with locally advanced breast cancer, the levels of SRC-1 and MYC were associated with each other and with reduced disease-free survival [Al-azawi et al., 2008]. Overall, the SRCs interact with and regulate several important cell cycle regulators whose functions are often aberrant in human cancers. The coactivators are found downstream in several of the central pathways regulating cell cycle progression and proliferation, but they also seem to modulate activation of growth factor signaling pathways involved in breast cancer establishment and progression, especially SRC-3/AIB1.

The SRCs have been shown to contribute to the process of metastasis, which is one of the hallmarks of cancer. In a mouse model of breast cancer, SRC-1 promoted metastatic behavior of the tumor cells by facilitating Ets-2 mediated HER-2/neu expression, and by increasing the expression of colony stimulating factor-1 (CSF-1) which recruits macrophages to mammary tumors [Wang et al., 2009]. In addition, it was shown *in vitro* using cells derived from mammary tumors that SRC-1 promotes migration and invasion by enhancing PEA3-mediated transcriptional activation of TWIST, a master regulator of metastasis [Qin et al., 2009]. Matrix metalloproteinases (MMPs) are zink-dependent enzymes involved in the degradation of extracellular matrix that is essential in the metastatic process. SRC-3/AIB1 has been suggested to promote breast cancer metastasis by stimulating the expression of various MMPs [Qin et al., 2008; Li et al., 2008; Yan et al., 2008]. SRC-1, SRC-2/TIF-2 and SRC-3/AIB1 have all been shown to increase the basal expression of SDF-1a in MCF-7-derived breast cancer cells. These cells secrete SDF-1a protein which regulates proliferation and invasion and which was found to be necessary for estrogen-induced breast cancer proliferation [Kishimoto et al., 2005].

Steroid Receptor Coactivators and Mechanism of Action in Endocrine Treatment

The important role of estrogen and ER in breast cancer development and progression has made endocrine therapy one of the central treatment modalities in ER positive breast cancer. Since the introduction of oophorectomy in the early 20th century, followed by adrenalectomy and hypophysectomy as a measure to halt tumor growth in breast cancer, it has become evident that hormones play a key role in breast cancer development and could be targeted therapeutically. SERMs, where tamoxifen is the most established drug, can now be used to block the estrogen receptor in both pre- and post-menopausal breast cancer whereas estrogen deprivation by the use of AIs can be offered to post-menopausal women with breast cancer (Figure 3).

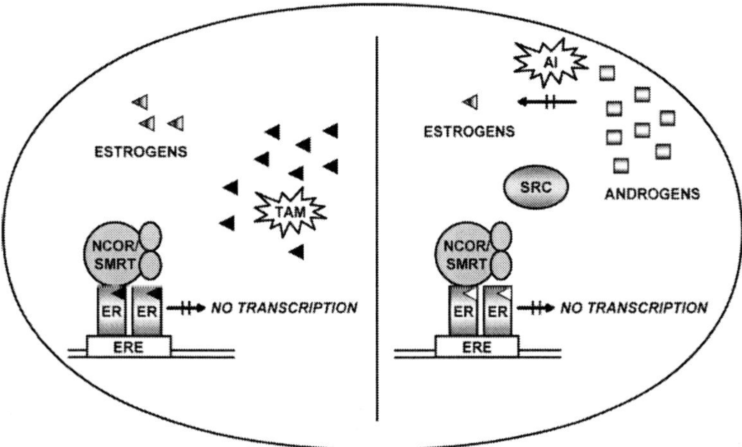

Figure 3. Schematic presentation of the main forms of endocrine treatment in breast cancer. Estrogen has growth stimulatory effects in breast cancer cells. In hormone receptor positive breast cancer, endocrine treatment aims at blocking the estrogen-stimulated proliferation of the malignant cells. Tamoxifen (tam) competes with estrogens for the ER binding site. Upon binding to ER, tamoxifen induces conformational changes in the ER leading to recruitment of corepressors and inhibition of ER regulated gene transcription at the estrogen response element (ERE) in the promoter of ER regulated genes. The aromatase inhibitors (AIs) inhibit estrogen synthesis and thereby inhibit the ligand-dependent activation of the ER.

The use of SERMs in breast cancer therapy has provided great insights into the regulatory mechanisms of ER transcriptional activity, which in turn has led to the development of new therapeutic possibilities, i.e. the use of SERMs such as raloxifene in osteoporosis. While the success of tamoxifen in breast cancer therapy is based on its ER antagonistic effects in malignant breast tissue, tamoxifen demonstrates ER agonistic effects in other organ systems such as bone and liver. SERMs typically act as ER agonist or antagonist in a tissue and cellular specific fashion. Results from cellular assays indicate that the SRCs are crucial for the SERM effect in any specific tissue. The expression levels of coactivators and corepressors, and the recruitment of coregulators to target gene promoters, can determine the effect of ER on gene expression. Experimental modulation of coactivator expression has been shown to influence the tamoxifen effect and alter ER transcriptional activity in response to endocrine treatment [Shang and Brown, 2002; Romano et al., 2010].

Upon binding to ER, SERMs typically inhibit ER transcriptional activity by competing with estradiol for the binding site and by blocking the AF-2 activity of ER [Shiau et al., 1998; Brzozowski et al., 1997]. AF-2 consists of surface exposed residues within helices 3, 5 and 12 in the LBD that form a hydrophobic pocket which binds the LXXLL-motifs in the NR-boxes of coactivators. The positioning of helix 12 after ligand binding to ER is the key event that regulates ER agonistic and antagonistic effects by dictating recruitment of transcriptional coactivators or corepressors [Pearce et al., 2003]. The potent ER antagonistic metabolite of tamoxifen, 4-hydroxytamoxifen, induces a displacement and rotation of the receptor's helix 12. The helix 12 then binds to the hydrophobic pocket via a sequence resembling the NR box of the coactivators, and thereby inhibits coactivator recruitment [Brzozowski et al., 1997; Shiau et al., 1998; Shiau et al., 2002]. The binding of 4-hydroxytamoxifen instead favors recruitment of two corepressors, silencing mediator for retinoid X receptor and thyroid hormone receptor (SMRT) and nuclear receptor corepressor (NCoR), resulting in inhibition of ER regulated gene transcription [Webb et al., 2003; Fleming et al., 2004].

While tamoxifen inhibits breast cancer tumor growth via antagonism of ER, tamoxifen may exert ER agonistic effects depending on the coactivator context. Studies of genes with classical EREs have shown that estrogenic effects of tamoxifen can be mediated by the constitutive active AF-1 domain of ER. The AF-1 domain is much less conserved between different nuclear receptors compared to AF-2, and can be stimulated by several mechanisms, including high levels of coactivators [Webb et al., 1998]. It has been shown in endometrial cancer cell lines and HepG2 cells that elevation of SRC-1 levels can lead to a shift from tamoxifen antagonistic to agonistic effects by recruitment of the coactivator to genes such as *c-MYC* and *IGF-1*. These genes do not have a classical ERE but are regulated by ER via indirect binding to the target gene promoter [Shang and Brown, 2002; Smith et al., 1997]. SRC-1 has also been shown to promote tamoxifen agonistic effects by bridging the transcriptional coactivator PGC-1β and the N-terminal domain of ER, resulting in increased transcriptional activity of the ER-tamoxifen complex [Kressler et al., 2007]. Moreover, tamoxifen acts as an ER agonist in breast cancer cell lines that have been engineered to overexpress SRC-3/AIB1 and the growth factor receptor HER-2/neu [Shou et al., 2004]. In line with this, *in vitro* studies have shown that dissociation of SRC-3/AIB1 from ER can inhibit breast cancer cell growth [Planas-Silva et al., 2001; List et al.,2001] and restore tamoxifen's ER antagonistic effect in resistant breast cancer cells [Wang et al., 2006]. Tamoxifen resistance with loss of ER antagonistic effects develops when SRC-3/AIB1 is high and the transcriptional repressor paired box 2 (PAX2) is low in breast cancer cells [Hurtado et al., 2008]. SRC-1 overexpression leading to enhanced tamoxifen-induced transcription can be inhibited by corepressors [Smith et al., 1997]. These data indicate that the level of SRCs in breast tumor tissue may determine the response to tamoxifen treatment.

It is also of note that AIs work by removing estrogen as the endogenous ligand of ER, thereby inhibiting ER transcriptional activity, regardless of whether it is regulated by the AF-1 or AF-2 domain. Thus, the mechanisms of AIs are clearly different from those of tamoxifen, which may explain differences in the clinical efficiency and resistance mechanism between different endocrine regimens.

Effects of Endocrine Treatment on Expression of Steroid Receptor Coactivators

The central role of the SRCs as mediators of treatment effects with endocrine therapy makes the regulation of coactivator expression during endocrine treatment especially interesting. Endocrine treatment in breast cancer would normally be continued for years, but the responsiveness to treatment can and often does change during the planned treatment period, due to development of endocrine resistance. Though changes in coactivator expression and activity may determine the sensitivity to endocrine therapy and induce changes in endocrine responsiveness over time, the expression levels of SRCs during estrogen deprivation are not well known. In a clinical study on neoadjuvant treatment with AIs in locally advanced breast cancer, we collected tumor samples from 30 ER positive breast cancer patients before and during 3 months of treatment with aromatase inhibitors, and measured mRNA levels of SRC-1, SRC-3/AIB1 and HER-2/neu by real-time RT-PCR [Flågeng et al., 2009]. Interestingly, the mRNA levels of coactivators and HER-2/neu increased in tumors during treatment, especially for SRC-1 (Figure 4). *In vitro* assays using the ER positive breast cancer cell line ZR75-1 are in line with our clinical data, showing that estrogen decreases the mRNA and protein level of HER-2/neu [Newman et al., 2000]. However, SRC-1 mRNA levels were found to be down-regulated in MCF-7 cells deprived of estrogen [Jeng et al., 1998].

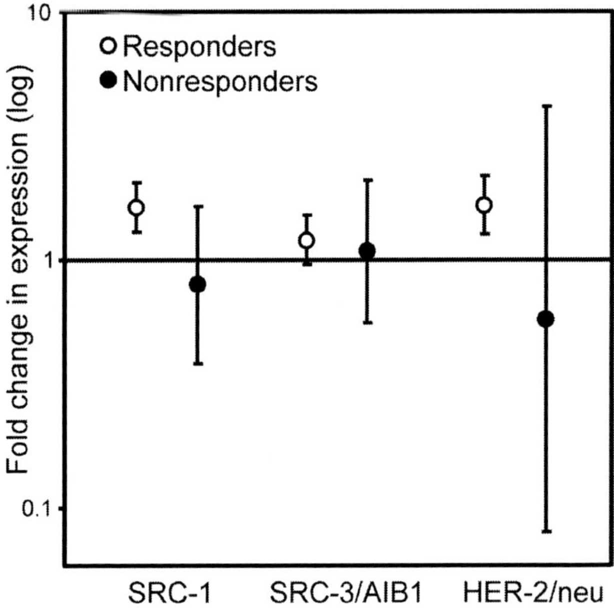

Figure 4. Effect of treatment with aromatase inhibitors on the mRNA expression of SRC-1, SRC-3/AIB1 and HER-2/neu. RNA was purified from tumors from breast cancer patients before and after 13–16 weeks of treatment with an aromatase inhibitor. Patients were classified as responders or nonresponders based on clinical tumor measurements during treatment. mRNA expression was estimated using real-time RT-PCR. Target gene concentrations were calculated relative to the housekeeping genes GAPDH and TBP. Fold change in mRNA expression during treatment are presented as geometric means with 95% confidence intervals. The mRNA expression of SRC-1 increased during treatment in the patient group as a whole ($P = 0.008$), as well as in the subgroup of patients achieving an objective treatment response ($P = 0.002$). No significant change in SRC-3/AIB1 level was recorded. There was an increase in HER-2/neu mRNA levels during therapy in the total patient group ($P = 0.016$) and in particular among responders ($P = 0.008$) [Flågeng et al., 2009].

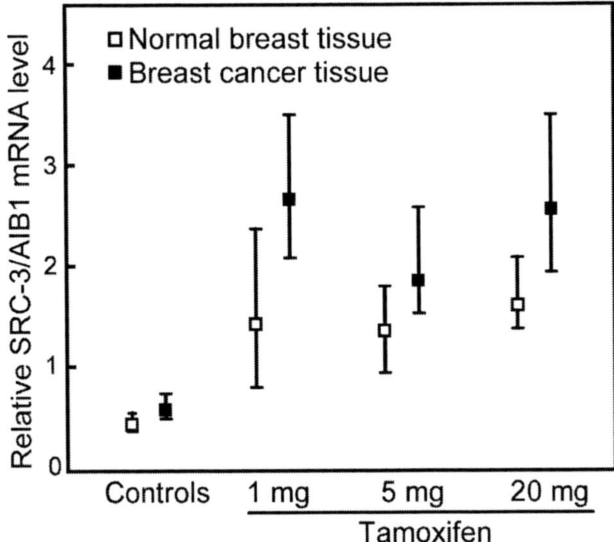

Figure 5. Effect of treatment with tamoxifen on the mRNA expression of SRC-3/AIB1 in normal and malignant breast tissue. Tumor and adjacent normal breast tissue specimens were collected at surgery after 4 weeks of preoperative tamoxifen treatment at doses of 1, 5 or 20 mg daily. mRNA expression levels were analyzed by real-time RT-PCR, calculated relative to the housekeeping gene TBP and are presented as geometric means with 95 % confidence intervals. Differences in mRNA levels between untreated and tamoxifen-treated tissue were statistically significant with $P < 0.001$ in normal and malignant breast tissue. The general pattern was the same for SRC-1 and SRC-2/TIF-2 [Haugan Moi et al., 2010].

We found a positive correlation between SRC-1 and HER-2/neu in human breast tissue treated with AIs. This finding is interesting in light of *in vitro* assays suggesting that ER and HER-2/neu compete for the coactivator SRC-1. Under antiestrogenic conditions, SRC-1 will be released from the ER and may instead bind to the HER-2/neu enhancer and facilitate transcription of HER-2/neu, leading to increased expression of HER-2/neu under estrogen-deprived conditions [Newman et al., 2000]. Upregulation of SRCs at the mRNA level might seem to be a general regulatory mechanism during early endocrine treatment. In line with *in vitro* studies we found that preoperative tamoxifen treatment for 4 weeks using tamoxifen doses from 1 to 20 mg/daily, significantly upregulated mRNA levels of all three SRCs in both normal and malignant breast tissue, especially for SRC-3/AIB1 [Haugan Moi et al., 2010] (Figure 5). Estradiol has been shown to repress SRC-3/AIB1 mRNA and protein expression in MCF-7 human breast cancer cells primarily by suppressing SRC-3/AIB1 gene transcription [Lauritsen et al., 2002].

Conversely, total SRC-3/AIB1 mRNA levels were increased when MCF-7 breast cancer cells were treated with the antiestrogen 4-hydroxytamoxifen. 4-hydroxtamoxifen has also been shown to increase the stability and hence steady-state levels of SRC-1 and SRC-3/AIB1 protein in an MCF-7 breast cancer-derived cell line [Lonard et al., 2004].

Steroid Receptor Coactivators and Response to Endocrine Treatment: Coactivators, ER and HER-2/Neu

There is accumulating evidence that SRC expression and activity are related to the clinical response to endocrine treatment and to disease-free survival in breast cancer. In our study on neoadjuvant treatment with aromatase inhibitors, the increase in SRC-1 mRNA levels during treatment was especially evident in tumors that responded to treatment [Flågeng et al., 2009] (Figure 6). On the other hand, the highest levels of SRC-3/AIB1 in breast tumors after 4 weeks of tamoxifen treatment were associated with unfavorable disease-free survival after a median follow-up time of 8 years [Haugan Moi et al., 2010]. The increase in SRC mRNA in endocrine sensitive tumors, but association between high SRC levels and worse long-term outcome might seem contradictory. Differences in study design, length of treatment, different medications and drug dose may have influenced our observations. The association between higher SRC-3/AIB1 levels after tamoxifen treatment, poor prognostic factors and decreased overall survival could theoretically also be attributed to other, unknown properties of the tumors with the highest SRC-3/AIB1 levels. However, it is also possible that an upregulation of the SRCs after introduction of an ER blocker or AIs signifies effective endocrine treatment in endocrine responsive cells. As described *in vitro*, loss of estrogen stimulation would lead to decreased transcription of the SRCs [Lonard et al., 2004; Lauritsen et al., 2002]. Increased mRNA levels of the SRCs reflecting endocrine responsiveness is also in line with a small study on recurrent breast cancer, demonstrating an association between higher SRC-1 mRNA levels in tumors and favorable response to tamoxifen treatment [Berns et al., 1998]. Elevated mRNA expression levels of SRC-3/AIB1 mRNA at surgery have been associated with shorter disease-free survival in the longer term [Zhao et al., 2003], which is in line with our study [Haugan Moi et al., 2010].

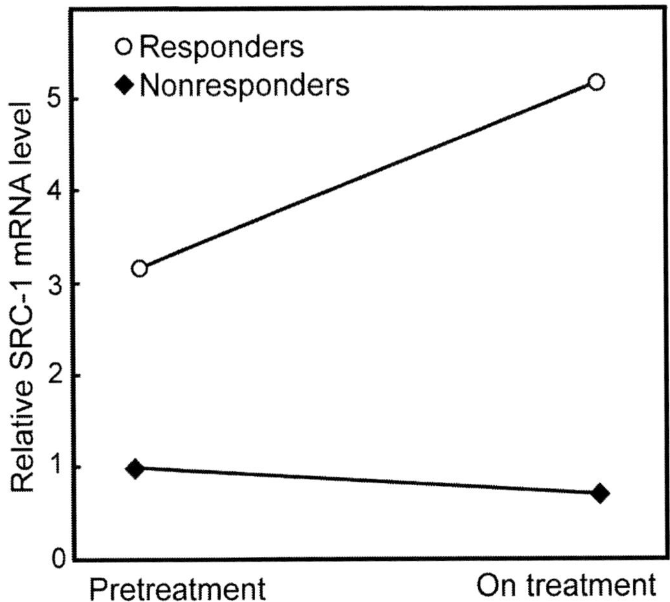

Figure 6. Change in SRC-1 mRNA expression during treatment with aromatase inhibitors in relation to clinical treatment response. Total RNA was extracted from tumors before and after 13-16 weeks of treatment with an aromatase inhibitor. SRC-1 mRNA expression was measured by real-time RT-PCR and calculated relative to the mRNA expression of the housekeeping genes GAPDH and TBP. Fold change values are presented as geometric means of SRC-1 mRNA expression before and on treatment for responders (n= 24) and nonresponders (n=6). The differences in the fold change of SRC-1 mRNA expression between responders and nonresponders were statistically significant ($P = 0.023$) [Flågeng et al., 2009].

The upregulation of the SRCs at the mRNA level may be interpreted as an early response to effective endocrine treatment in estrogen responsive breast cancer cells. However, post-translational modifications of the SRCs via activation of signaling pathways (e.g. growth factors) could contribute to functional activation of the coactivators with detrimental effects on endocrine responsiveness. Indeed, it has been shown that phosphorylation is especially important for regulation of SRC-3/AIB1 mediated activity on steroid and growth factor signaling and cell transformation [Font de Mora and Brown, 2000; Wu et al., 2002; Wu et al., 2004].

Interestingly, several clinical studies have indicated that the coactivators have prognostic power and are associated with endocrine response and disease-free survival in ER positive breast cancer. High levels of SRC-1 protein were found to associate with reduced disease-free survival, both in untreated and tamoxifen treated patients [Al-Azawi et al., 2007; Redmond et al., 2009]. Conversely, low SRC-1 expression levels combined with high ERβ expression were shown to serve as favorable prognostic indicators [Myers et al., 2004]. High levels of SRC-3/AIB1 mRNA levels have been associated with high tumor grade, lack of protein staining for ER and progesterone receptor and high protein levels of p53 and HER-2/neu [Bouras et al., 2001]. SRC-3/AIB was found to be a positive prognostic factor in untreated patients [Osborne et al., 2003], but associated with worse outcome in another study on an unselected cohort of ER negative breast cancer patients [Harigopal et al., 2009]. High SRC-3/AIB1 protein levels have been associated with reduced disease-free survival in tamoxifen treated samples in several studies [Myers et al., 2005; Dihge et al., 2007], if not all [Alkner et al., 2010]. However, the intracellular localization of the coactivator may also be of relevance, since SRC-3/AIB1 nuclear protein staining has been associated with improved response to endocrine treatment [Iwase et al., 2003; Alkner et al., 2010].

SRCs and HER-2/neu

Several studies have shown that overexpression of SRC-3/AIB1 in invasive breast cancer correlates with high levels of HER-2/neu [Bouras et al., 2001; Osborne et al., 2003]. HER-2/neu belongs to the family of epidermal growth factor receptors and is overexpressed in >30% of breast cancers. High levels of HER-2/neu associate with negative prognosis and reduced disease-free survival [Slamon et al., 1987]. The HER-2/neu protein can be targeted in HER-2/neu positive breast cancers using the monoclonal antibody trastuzumab. The clearest and most consistent findings of an association between SRCs, endocrine responsiveness and long-term outcome have been shown in studies investigating SRC levels and HER-2/neu expression in relation to therapy response. Patients undergoing tamoxifen therapy with tumors overexpressing HER-2/neu in combination with SRC-3/AIB1 or SRC-1 show reduced sensitivity to endocrine treatment, greater probability of disease recurrence and reduced disease-free survival [Osborne et al., 2003; Fleming et al., 2004; Kirkegaard et al., 2007]. Experiments on cell lines have also demonstrated that overexpression of HER-2/neu in ER positive cells can result in resistance to tamoxifen [Pietras et al.,1995]. Tamoxifen behaves like an estrogen agonist in ER-positive breast cancer cells that express high levels of SRC-3/AIB and HER-2/neu, resulting in *de novo* resistance [Shou et al., 2004].

These studies suggest that putative interactions between SRCs and HER-2/neu, or other growth factors, may determine the treatment response and long-term outcome in ER positive breast cancer. The association between HER-2/neu, coactivators and endocrine responsiveness is especially interesting since the activity of the tamoxifen-ER complex can be modulated by phosphorylation of ER and/or coactivators by kinases found downstream of HER-2/neu [Feng et al., 2001]. HER-2/neu activates mitogen activated kinases (MAPKs). Both SRC-1 and SRC-3/AIB1 are phosphorylated and transcriptionally activated by MAPKs that stimulate the recruitment of the cointegrator CBP/p300 and enhance the histone acetyltransferase activity of the SRCs *in vitro* [Rowan et al., 2000; Font de Mora and Brown, 2000]. The SRCs are recruited to the ER in presence of tamoxifen and an activated HER-2/neu-MAPK system [McIlroy et al., 2006], which could lead to tamoxifen resistance [Benz et al., 1992; Miller et al., 1994]. Experiments have shown that SRC-3/AIB1 and HER-2/neu signaling were involved in the proliferation of tamoxifen resistant breast cancer cells. Silencing of SRC-3/AIB1 with siRNA significantly reduced the HER-2/neu stimulated cell growth, and tamoxifen sensitivity was restored [Zhao et al., 2009]. It has also been suggested from *in vitro* assays that the role of SRC-1 in regulating SDF-1a expression and of HER-2/neu in stabilizing the SDF-1a receptor CXCR4 may partly explain the poor prognosis in tumors that overexpress both SRC-1 and HER-2/neu [Kishimoto et al., 2005].

HER-2/neu activates kinases that can modify the activity of the SRCs, but HER-2/neu is also regulated by the SRCs. Estrogen down-regulates HER-2/neu mRNA and protein levels in ER positive breast epithelial cells [Dati et al., 1990; Read et al., 1990, Warri et al., 1991, Antoniotti et al., 1994]. However, this effect is reversed in the presence of tamoxifen where SRC-1 released from the tamoxifen-ER complex has been shown to stimulate HER-2/neu transcription [Newman et al., 2000]. It has been suggested by *in vitro* studies that the MAPK dependent transcription factor Ets-1, which is found downstream of HER-2/ncu, may regulate the transcription of HER-2/neu through interaction with SRC-1 [Myers et al., 2005]. It is tempting to speculate that the reciprocal positive regulation of HER-2/neu and SRCs may explain the detrimental effects of overexpressed HER-2/neu and SRC-1 and/or SRC-3/AIB1 on endocrine responsiveness and disease-free survival, as observed by association in several studies of ER-positive breast cancer. Such a positive feedback loop with increased expression of SRC-3/AIB1 in HER-2/neu induced tumors, leading to further activation of HER-2/neu in breast tumor tissue and not in normal adjacent mammary tumors, has actually been observed *in vivo* in a mouse model of HER-2/neu induced breast cancer [Fereshteh et al., 2008]. In this mouse model, homozygous deletion of SRC-3/AIB1 led to loss of HER-2/neu induced tumor formation. Deletion of one allele of SRC-3/AIB1 delayed HER-2/neu-induced tumor formation and significantly decreased levels of phosphorylated HER-2/neu, cyclin D1, Akt and c-Jun NH2 kinase activation and proliferation in the malignant breast tumors. These observations would suggest that SRC-3/AIB1 is required for HER-2/neu oncogenic activity, at least in this mouse model system.

Overexpression of HER-2/neu also seems to be of special relevance to the development of resistance to AIs. MCF-7 cells stably expressing cellular aromatase (MCF-7/CA) and overexpressing HER-2/neu (MCF-7/CA/HER-2) show increased association of SRC-3/AIB1 and CBP with the estrogen target gene pS2 promoter compared to control MCF-7/CA cells upon treatment with androstendione. MCF-7/CA/HER-2 cells also show increased proliferation compared to MCF-7/CA cells in charcoal-stripped serum, and treatment with AIs did not abrogate cell proliferation of MVF-7/CA/HER-2 cells [Shin et al., 2006]. These observations suggest that ligand-independent recruitment of coactivator complexes to estrogen-responsive promoters, as a result of HER-2/neu overexpression, may play a role in development of resistance, also to AIs.

SRCs and MAPK Signaling

The MAPKs found downstream of HER-2/neu are also activated in the frequently used cellular model system for resistance to AIs, the long-term estrogen deprived cells (LTED). Breast cancer cells grown in estrogen-deprived conditions for 1-6 months develop enhanced sensitivity to estradiol [Masamura et al., 1995; Santen et al., 2005]. This hypersensitivity is associated with upregulation of ERα and the MAPKs [Jeng et al., 1998; Jeng et al., 2000]. However, it is important to underline that the MAPKs are found downstream of several growth factor receptors that could functionally activate not only the SRCs, but also the ER- and other pathways. Based on data from cellular assays, post-translational modification of both coactivators and ER can lead to altered molecular conformations, functional activation, increased degradation or intracellular compartmentalization, all of which would influence the effect of endocrine treatment on ER transcriptional activity. The MAPKs have been shown to phosphorylate not only the SRCs but also ER [Bunone et al., 1996; Kato et al., 1995; Lannigan 2003], and Akt also phosphorylates both ER, SRC-2/TIF-2 and SRC-3/AIB1 *in vitro* [Sun et al., 2001; Font and Brown, 2000; Lopez et al., 2001]. In endometrial cells the protooncogene Src kinase, which is deregulated in many cancers, has been shown to increase the transcriptional activity of SRC-1 [Shah and Rowan, 2005].

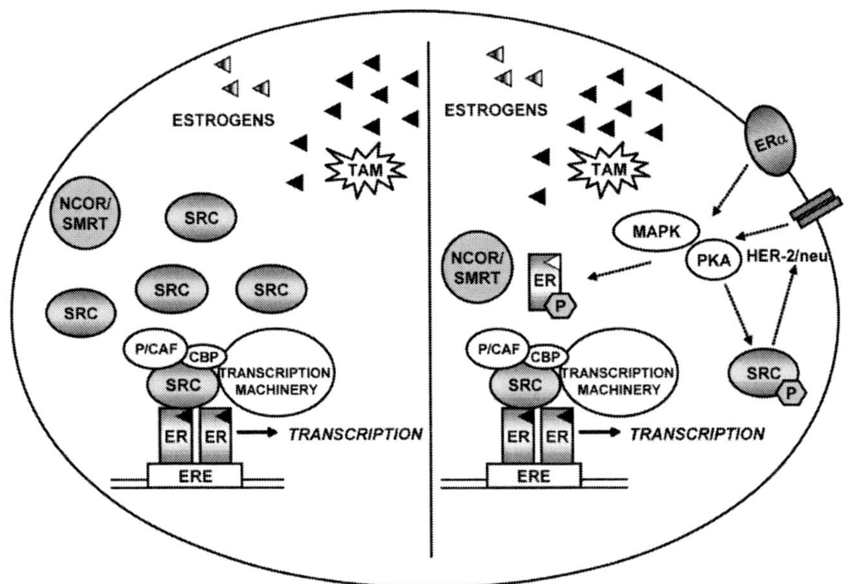

Figure 7. Some suggested mechanisms involved in ER agonistic effects of tamoxifen. High levels of steroid receptor coactivators (SRCs) relative to corepressors such as SMRT and NCoR may force ER into an active structural conformation. Recruitment of coactivators to the ER-tamoxifen complex leads to transcription of ER regulated genes and ER agonistic effects mediated by tamoxifen. *In vitro* and clinical studies indicate an association between HER-2/neu expression, high levels of SRC and ER agonistic effects of tamoxifen contributing to endocrine resistance. Kinases such as MAPK and PKA can be activated by growth factor signaling pathways and are found downstream of HER-2/neu and membrane-bound ER, among others. They can phosphorylate and functionally activate SRCs and/or ER whereas SRCs have been shown to positively regulate the expression levels and activity of HER-2/neu, so that a positive feedback loop can exist between HER-2/neu and SRC expression and activity. These mechanisms can theorctically be involved in ER agonistic effects of tamoxifen and contribute to endocrine resistance. It is underlined that the illustration is highly simplified.

Phosphorylation of SRC-1 by protein kinase A (PKA) has been shown to enhance ligand-independent activation of a nuclear receptor [Rowan et al., 2000; Carascossa et al., 2010]. Also, we have reported that PKA specifically induces ubiquitination and degradation of SRC-2/TIF-2 [Hoang et al., 2004] and stimulates recruitment of SRC-2/TIF-2 to ER [Fenne et al., 2008]. In summary, the steroid receptor coactivators and ER have been shown to interact with growth factor receptor pathways, which may contribute to cancer cell proliferation and invasion in the absence of estrogen and influence the response to endocrine treatment (Figure 7). The coactivators seem to play a pivotal role in the crosstalk between ER and growth factor pathways.

FUTURE PERSPECTIVES: SRCS AS THERAPEUTIC TARGETS?

In vitro assays have indicated that the relative level of activity of coactivators to corepressors may dictate the effect of SERMs in different tissues, and influence the activity of ER during estrogen deprivation. The pivotal role of the SRCs in ER regulated gene transcription and ER-HER-2/neu crosstalk is supported by the association of SRCs and HER-2/neu co-overexpression with response to endocrine treatment and disease-free survival. The ability of the SRCs, especially SRC-3/AIB1, to enhance the activity of multiple signal transduction pathways involved in cancer, including HER-2/neu, IGF-I and Akt, supports the possibility that SRC-3/AIB1 could not only be a predictive marker for cancer therapy, but could also be a potential therapeutical target. Downregulation of SRC-3/AIB1 may decrease the crosstalk between ER and the HER-2/neu pathway, and has been shown to restore tamoxifen sensitivity in tamoxifen-resistant breast cancer cells. Theoretically, there are several options available if one wants to target SRC-3/AIB1 in cancer cells, such as reducing the levels of SRC-3/AIB1, targeting kinases that functionally activate the coactivator or inhibiting the coactivator's function by disrupting its interaction with CBP/p300. Provided that increased levels of HER-2/neu and the steroid receptor coactivators during endocrine treatment are important mechanisms for the development of resistance to tamoxifen and AIs over time, new treatment regimes with existing treatment modalities such as targeting the HER-2/neu pathway might be beneficial to selected patient subgroups. However, these patient populations need to be identified through thorough clinical studies. Therapies that target not only ER and HER-2/neu, but also the kinases found downstream of HER-2/neu or other growth factor signaling pathways regulated by the SRCs and involved in the crosstalk between ER, HER-2/neu and the SRCs, such as IGF-1 or Akt, are interesting as therapeutical targets in selected patient groups.

Most of our knowledge of the transcriptional regulation of ER during endocrine treatment comes from experiments in different cancer cell lines and with a limited number of promoters investigated. Studies of regulatory mechanisms *in vivo* are starting to be revealed. The continuous research on gene expression profiles and protein expression in human tumors at diagnosis, during treatment and at the time of treatment failure combined with data on treatment response and long-term outcome will hopefully result in important contributions to our understanding of human breast cancer. Insight into the regulatory mechanisms involved in endocrine therapy and in the development of treatment resistance *in vivo* brings hope of further therapeutic improvements to the many women being diagnosed with breast cancer.

CONCLUSION

In summary, the SRCs are expressed in normal and malignant breast tissue and they have a crucial role in ER regulated gene transcription. SRC-3/AIB1 has been shown to have oncogenic properties, and both SRC-1 and SRC-3/AIB1 can contribute to the metastatic process in breast cancer. As regulators of ER transcriptional activity, the SRCs are important mediators of the effect of endocrine treatment in breast cancer, and have been pointed out as the most important determinants of SERM effects in different tissues. Interestingly, the expression levels of SRCs are regulated by endocrine treatment with an upregulation at the mRNA level during antiestrogen treatment *in vitro* and *in vivo*. Importantly, the functional role of the SRCs is modified by post-translational modifications mediated by growth factor pathways that are involved in breast cancer development and endocrine resistance, such as HER-2/neu, with increasing evidence of a positive feedback loop between HER-2/neu activity and SRC expression in breast cancer. Clinically, several studies suggest that the concentration of coactivators influences the response to endocrine treatment, particularly in patients with tumors overexpressing HER-2/neu. Expanding our knowledge of steroid receptor coactivators, their interaction with ER and HER-2/neu, and their involvement in crosstalk between different growth factor signaling pathways under endocrine treatment, will be a central focus of future research on breast cancer therapy and endocrine resistance. Future research may reveal a role for the coactivators as direct or indirect therapeutical targets, also in endocrine resistant breast cancer.

REFERENCES

Al-azawi, D., McIlroy, M., Kelly, G., Redmond, A.M., Bane, F.T., Cocchiglia, S., Hill, A.D.K. and Young, L.S. (2008). Ets-2 and p160 proteins collaborate to regulate c-Myc in endocrine resistant breast cancer. *Oncogene*, 27, 3021-3031.

Alkner, S., Bendahl, P.O., Grabau, D., Lövgren, K., Stål, O., Rydén, L., Fernö, M.; Swedish and South-East Swedish Breast Cancer Groups. (2010). AIB1 is a predictive factor for tamoxifen response in premenopausal women. *Ann. Oncol.*, 21, 238-244.

Antoniotti, S., Tavema, D., Maggiora, P., Sapei, M.L., Hynes, N.E. and DeBortoli, M. (1994). Oestrogen and epidermal growth factor down-regulate erbB-2 oncogene protein expression in breast cancer cells by different mechanisms. *Br. J. Cancer*, 70, 1095-1101.

Anzick, S.L., Kononen, J., Walker, R.L., Azorsa, D.O., Tanner, M.M., Guan, X.Y., Sauter, G., Kallioniemi, O.P., Trent, J.M. and Meltzer, P.S. (1997). AIB1, a steroid receptor coactivator amplified in breast and ovarian cancer. *Science*, 277, 965-968.

Bautista, S., Valles, H., Walker, R.L., Anzick, S.L., Zeillinger, R., Meltzer, P.S. and Theillet, C. (1998). In breast cancer, amplification of the steroid receptor coactivator gene AIB1 is correlated with estrogen and progesterone receptor positivity. *Clin. Cancer Res.*, 4, 2925–2929.

Benz, C.C., Scott, G.K., Sarup, J.C., Johnson, R.M., Tripathy, D., Coronado, E., Shepard, H.M. and Osborne, C.K. (1992). Estrogen-dependent, tamoxifen-resistant tumorigenic growth of MCF-7 cells transfected with HER2/neu. *Breast Cancer Res. Treat*, 24, 85-95.

Berns, E.M., van Staveren, I.L., Klijn, J.G. and Foekens, J.A. (1998). Predictive value of SRC-1 for tamoxifen response of recurrent breast cancer. *Breast Cancer Res. Treat*, 48, 87-92.

Bouras T., Southey, M.C. and Venter, D.J. (2001). Overexpression of the steroid receptor coactivator AIB1 in breast cancer correlates with the absence of estrogen and progesterone receptors and positivity for p53 and HER2/neu. *Cancer Res.*, 61, 903-907.

Brzozowski, A.M., Pike, A.C., Dauter, Z., Hubbard, R.E., Bonn, T., Engstrom, O., Ohman, L., Greene, G.L., Gustafsson, J.A. and Carlquist, M. (1997). Molecular basis of agonism and antagonism in the oestrogen receptor. *Nature*, 389, 753-758.

Bunone, G., Briand, A., Miksicek, R.J. and Picard, D. (1996). Activation of the unliganded estrogen receptor by the EGF involves the MAP kinase pathway and direct phosphorylation. *EMBO J.*, 15, 2174-2183.

Carapeti, M., Agular, R.C., Chase, A., Goldman, J.M. and Cross, N.C. (1998). Assignment of the steroid receptor coactivator-1 (SRC-1) gene to human chromosome band 2p23. *Genomics*, 52, 242-244.

Carascossa, S., Dudek, P., Cenni, B., Briand, P.A. and Picard, D. (2010). CARM1 mediates the ligand-independent and tamoxifen-resistant activation of the estrogen receptor alpha by cAMP. *Genes Dev.*, 24, 708-719.

Chauchereau, A., Amazit, L., Quesne, M., Guiochon-Mantel, A. and Milgrom, E. (2003). Sumoylation of the progesterone receptor and of the steroid receptor coactivator SRC-1. *J. Biol. Chem.*, 278, 12335-12343.

Cuzick, J., Powles, T., Veronesi, U., Forbes, J., Edwards, R., Ashley, S. and Boyle P. (2003). Overview of the main outcomes in breast-cancer prevention trials. *Lancet*, 361, 296-300.

Dati, C., Antoniotti, S., Tavema, D., Perroteau, I. and De Bortoli, M. (1990). Inhibition of c-erbB-2 oncogene expression by estrogens in human breast cancer cells. *Oncogene*, 5, 1001-1006.

Dihge, L., Bendahl, P.O., Grabau, D., Isola, J., Lövgren, K., Rydén, L. and Fernö, M. (2008). Epidermal growth factor receptor (EGFR) and the estrogen receptor modulator amplified in breast cancer (AIB1) for predicting clinical outcome after adjuvant tamoxifen in breast cancer. *Breast Cancer Res. Treat*, 109, 255-262.

DiRenzo, J., Shang, Y., Phelan, M., Sif, S., Myers, M., Kingston, R. and Brown M. (2000). BRG-1 is recruited to estrogen-responsive promoters and cooperates with factors involved in histone acetylation. *Mol. Cell Biol.*, 20, 7541-7549.

Dubik, D. and Shiu, R.P. (1992). Mechanism of estrogen activation of c-myc oncogene expression. *Oncogene*, 7, 1587-1594.

Early Breast Cancer Trialists' Collaborative Group. (2005). Effects of chemotherapy and hormonal therapy for early breast cancer on recurrence and 15-year survival: an overview of the randomised trials. *Lancet*, 365, 1687-1717.

Feng, W., Webb, P., Nguyen, P., Liu, X., Li, J., Karin, M. and Kushner, P.J. (2001). Potentiation of estrogen receptor activation function 1 (AF-1) by Src/JNK through a serine 118-independent pathway. *Mol. Endocrinol.*, 15, 32-45.

Fenne, I.S., Hoang, T., Hauglid, M., Sagen, J.V., Lien, E.A. and Mellgren, G. (2008). Recruitment of coactivator glucocorticoid receptor interacting protein 1 to an estrogen receptor transcription complex is regulated by the 3',5'-cyclic adenosine 5'-monophosphate-dependent protein kinase. *Endocrinology*, 149, 4336-4345.

Fereshteh, M.P., Tilli, M.T., Kim, S.E., Xu, J., O'Malley, B.W., Wellstein, A., Furth, P.A. and Riegel, A.T. (2008). The nuclear receptor coactivator amplified in breast cancer-1 is required for Neu (ErbB2/HER2) activation, signaling, and mammary tumorigenesis in mice. *Cancer Res.*, 68, 3697-3706.

Fleming, F.J., Hill, A.D.K., McDermott, E.W., O'Higgins, N.J. and Young, L.S. (2004). Differential recruitment of coregulator proteins steroid receptor coactivator-1 and silencing mediator for retinoid and thyroid receptors to the estrogen receptor-estrogen response element by b-estradiol and 4-hydroxytamoxifen in human breast cancer. *J. Clin. Endocrinol. Metab.*, 89, 375-383.

Fleming, F.J., Myers, E., Kelly, G., Crotty, T.B., McDermott, E.W., O'Higgins, N.J., Hill, A.D.K. and Young, L.S. (2004). Expression of SRC-1, AIB1, and PEA3 in HER2 mediated endocrine resistant breast cancer; a predictive role for SRC-1. *J. Clin. Pathol.*, 57, 1069-1074.

Flågeng, M.H., Haugan Moi, L.L., Dixon, J.M., Geisler, J., Lien, E.A., Miller, W.R., Lønning, P.E. and Mellgren G. (2009). Nuclear receptor co-activators and HER-2/neu are upregulated in breast cancer patients during neo-adjuvant treatment with aromatase inhibitors. *Br. J. Cancer*, 101, 1253-1260.

Font de Mora, J. and Brown, M. (2000). AIB1 is a conduit for kinase-mediated growth factor signaling to the estrogen receptor. *Mol. Cell Biol.*, 20, 5041-5047.

Foster, J.S., Henley, D.C., Ahamed, S. and Wimalasena, J. (2001). Estrogens and cell-cycle regulation in breast cancer. *Trends Endocrinol Metab*, 12, 320-327.

Gojis, O., Rudraraju, B., Alifrangis, C., Krell, J., Libalova, P. and Palmieri, C. (2009). The role of steroid receptor coactivator-3 (SRC-3) in human malignant disease. *Eur. J. Surg. Oncol.*, 36, 224-229.

Goldhirsch, A., Ingle, J.N., Gelber, R.D., Coates, A.S., Thürlimann, B., Senn, H.-J.and Panel members. (2009). Thresholds for therapies: highlights of the St Gallen International Expert Consensus on the Primary Therapy of Early Breast Cancer 2009. *Ann. Oncol.*, 20, 1319-1329.

Harigopal, M., Heymann, J., Ghosh, S., Anagnostou, V., Camp, R.L. and Rimm, D.L. (2009). Estrogen receptor co-activator (AIB1) protein expression by automated quantitative analysis (AQUA) in a breast cancer tissue microarray and association with patient outcome. *Breast Cancer Res. Treat,* 115, 77-85.

Haugan Moi, L.L., Flågeng, M.H., Gandini, S., Guerrieri-Gonzaga, A., Bonanni, B., Lazzeroni, M., Gjerde, J., Lien, E.A., DeCensi, A. and Mellgren, G. (2010). Effect of low dose tamoxifen on Steroid Receptor Coactivator 3/Amplified in Breast Cancer 1 in normal and malignant human breast tissue. *Clin. Cancer Res.*, 16, 2176-2186.

Hoang, T., Fenne, I.S., Cook, C., Børud, B., Bakke, M., Lien, E.A. and Mellgren, G. (2004). cAMP-dependent protein kinase regulates ubiquitin-proteasome-mediated degradation and subcellular localization of the nuclear receptor coactivator GRIP1. *J. Biol. Chem.* 279, 49120-49130.

Hudelist, G., Czerwenka, K., Kubista, E., Marton, E., Pischinger, K. and Singer, C.F. (2003). Expression of sex steroid receptors and their co-factors in normal and malignant breast tissue: AIB1 is a carcinoma-specific co-activator. *Breast Cancer Res. Treat*, 78, 193-204.

Hurtado, A., Holmes, K.A., Geistlinger, T.R., Hutcheson, I.R., Nicholson, R.I., Brown, M., Jiang, J., Howat, W.J., Ali, S. and Carroll, J.S. (2008). Regulation of ERBB2 by oestrogen receptor-PAX2 determines response to tamoxifen. *Nature*, 456, 663-666.

Iwase, H., Omoto, Y., Toyama, T., Yamashita, H., Hara, Y., Sugiura, H. and Zhang, Z. (2003). Clinical significance of AIB1 expression in human breast cancer. *Breast Cancer Res. Treat*, 80, 339-345.

Jeng, M.H., Shupnik, M.A., Bender, T.P., Westin, E.H., Bandyopadhyay, D., Kumar, R., Masamura, S. and Santen, R.J. (1998). Estrogen receptor expression and function in long-term estrogen-deprived human breast cancer cells. *Endocrinology*, 139, 4164-4174.

Jeng, M.H., Yue ,W., Eischeid, A., Wang, J.P. and Santen, R.J. (2000). Role of MAP kinase in the enhanced cell proliferation of long term estrogen deprived human breast cancer cells. *Breast Cancer Res. Treat*, 62, 167-175.

Kalkhoven, E., Valentine, J.E., Heery, D.M. and Parker, M.G. (1998). Isoforms of steroid receptor co-activator 1 differ in their ability to potentiate transcription by the oestrogen receptor. *EMBO J.*, 17, 232-243.

Karmakar, S., Foster, E.A. and Smith, C.L. (2009). Unique roles of p160 coactivators for regulation of breast cancer cell proliferation and estrogen receptor-alpha transcriptional activity. *Endocrinology*, 150, 1588-1596.

Kato, S., Endoh, H., Masuhiro, Y., Kitamoto, T., Uchiyama, S., Sasaki, H., Masushige, S., Gotoh, Y., Nishida, E. and Kawashima, H. (1995). Activation of the estrogen receptor through phosphorylation by mitogen-activated protein kinase. *Science*, 270, 1491-1494.

Kelly, M.J. and Levin, E.R. (2001). Rapid actions of plasma membrane estrogen receptors. *Trend Endocrinol. Metab.*, 12, 152-156.

Kershah, S.M., Desouki, M.M., Koterba, K.L. and Rowan, B.G. (2004). Expression of estrogen receptor coregulators in normal and malignant human endometrium. *Gynecol. Oncol.*, 92, 304-313.

Kirkegaard, T., McGlynn, L.M., Campbell ,F.M., Müller, F.M., Tovey, S.M., Dunne, B., Nielsen, K.V., Cooke, T.G. and Bartlett, J.M.S. (2007). Amplified in breast cancer 1 in human epidermal growth factor receptor – positive tumors of tamoxifen-treated breast cancer patients. *Clin. Cancer Res*, 13, 1405-1411.

Kishimoto, H., Wang, Z., Bhat-Nakshatri, P., Chang, D., Clarke, R. and Nakshatri, H. (2005). The p160 family coactivators regulate breast cancer proliferation and invasion through autocrine/paracrine activity of SDF-1A/CXCL12. *Carcinogenesis*, 26, 1706-1715.

Kressler, D., Hock, M.B. and Kralli A. (2007). Coactivators PGC-1b and SRC-1 interact functionally to promote the agonist activity of the selective estrogen receptor modulator tamoxifen. *J. Biol. Chem.*, 282, 26897-26907.

Ku, J., Qiu, Y., DeMayo, F.J., Tsai, S.Y., Tsai, M.J. and O'Malley, B.W. (1998). Partial hormone resistance in mice with disruption of the steroid receptor coactivator-1 (SRC-1) gene. *Science*, 279, 1922-1925.

Kuang, S.-Q., Liao, L., Wang, S., Medina, D., O'Malley, B.W. and Xu, J. (2002). Mice lacking the amplified in breast cancer 1/Steroid receptor coactivator-3 are resistant to chemical carcinogen-induced mammary tumorigenesis. *Cancer Res.*, 65, 7993-8002.

Kurebayashi, J., Otsuki, T., Kunisue, H., Tanaka, K., Yamamoto, S. and Sonoo, H. (2000). Expression levels of estrogen receptor-a, estrogen receptor-b, coactivators and corepressors in breast cancer. *Clin. Cancer Res.*, 6, 512–518.

Lahusen, T., Henke, R.T., Kagan, B.L., Wellstein, A. and Riegel, A.T. (2009). The role and regulation of the nuclear receptor co-activator AIB1 in breast cancer. *Breast Cancer Res. Treat*, 116, 225-237.

Lannigan, D.A. (2003). Estrogen receptor phosphorylation. *Steroids*, 68, 1-9.

Lauritsen, K.J., List, H.J., Reiter, R., Wellstein, A. and Riegel, A.T. (2002). A role for TGF-beta in estrogen and retinoid mediated regulation of the nuclear receptor coactivator AIB1 in MCF-7 breast cancer cells. *Oncogene*, 21, 7147-7155.

Lee, S.K., Kim, H.J., Na, S.Y., Kim, T.S., Choi, H.S., Im, S.Y. and Lee JW. (1998). Steroid receptor coactivator-1 coactivates activating protein-1-mediated transactivations through interaction with the c-Jun and c-Fos subunits. *J. Biol. Chem*, 273, 16651-16654.

Li, L.B., Louie, M.C., Chen, H.-W. and Zou, J.X. (2008). Proto-oncogene ACTR/AIB1 promotes cancer cell invasion by up-regulating specific matrix metalloproteinase expression. *Cancer Letters*, 261, 64-73.

Lien, E.A., Solheim, E., Lea, O.A., Lundgren, S., Kvinnsland, S. and Ueland, P.M. (1989). Distribution of 4-hydroxy-N-desmethyltamoxifen and other tamoxifen metabolites in human biological fluids during tamoxifen treatment. *Cancer Res.*, 49, 2175–2183.

Lippman, M. and Bolan, G. (1975). Oestrogen responsive human breast cancer in long term tissue culture. *Nature*, 256, 592–593.

List, H.J., Lauritsen, K.J., Reiter, R., Powers, C., Wellstein, A. and Riegel, A.T. (2001). Ribozyme targeting demonstrates that the nuclear receptor coactivator AIB1 is a rate-limiting factor for estrogen-dependent growth of human MCF-7 breast cancer cells. *J. Biol. Chem*, 276, 23763-23768.

List, H.J., Reiter, R., Singh, B., Wellstein, A. and Riegel, A.T. (2001). Expression of the nuclear coactivator AIB1 in normal and malignant breast tissue. *Breast Cancer Res. Treat*, 68, 21-28.

Lonard, D.M., Tsai, S.Y. and O'Malley, B.W. (2004). Selective estrogen receptor modulators 4-hydroxytamoxifen and raloxifene impact the stability and function of SRC-1 and SRC-3 coactivator proteins. *Mol. Cell Biol*, 24, 14-24.

Lonard, D.M., Kumar, R. and O'Malley, B.W. (2010). Minireview: the SRC family of coactivators: an entrée to understanding a subset of polygenic diseases? *Mol. Endocrinol*, 24, 279-285.

Lopez, G.N., Turck, C.W., Schaufele, F., Stallcup, M.R. and Kushner, P.J. (2001). Growth factors signal to steroid receptors through mitogen-activated protein kinase regulation of p160 coactivator activity. *J. Biol. Chem*, 276, 22177-22182.

Lukk, M., Kapushesky, M., Nikkilä, J., Parkinson, H., Goncalves, A., Huber, W., Ukkonen, E. and Brazma, A. Human gene expression atlas of 5372 samples representing 369 different cell and tissue types, disease states and cell lines [online]. 2010 [cited 27.09.10]. Available from: http://www.ebi.ac.uk/gxa/experiment/E-MTAB-62.

Lydon, J.P. and O'Malley, B.W. (2011). Minireview: steroid receptor coactivator-3: a multifarious coregulator in mammary gland metastasis. *Endocrinology*, 152, 19-25.

Mak, H.Y., Hoare, S., Henttu, P.M. and Parker, M.G. (1998). Molecular determinants of the estrogen receptor-coactivator interface. *Mol. Cell Biol.*, 19, 3895-3903.

Masamura, S., Santner, S.J., Heitjan, D.F. and Santen, R.J. (1995). Estrogen deprivation causes estradiol hypersensitivity in human breast cancer cells. *J. Clin. Endocrinol. Metab*, 80, 2918-2925.

McIlroy, M., Fleming, F.J., Buggy, Y., Hill, A.D.K. and Young, L.S. (2006). Tamoxifen induces ER-alpha – SRC-3 interaction in HER2 positive human breast cancer; a possible mechanism for ER isoform specific recurrence. *Endocr. Relat. Cancer*, 13, 1135-1145.

McNeil, C.M., Sergio, C.M., Anderson, L.R., Inman, C.K., Eggleton, S.A., Murphy, N.C., Millar, E.K., Crea, P., Kench, J.G., Alles, M.C., Gardiner-

Garden, M., Ormandy, C.J., Butt, A.J., Henshall, S.M., Musgrove, E.A. and Sutherland, R.L. (2006). c-Myc overexpression and endocrine resistance in breast cancer. *J. Steroid. Biochem. Mol. Biol*, 102, 147-55.

Miller, D.L., el-Ashry, D., Cheville, A.L., Liu, Y., McLeskey, S.W. and Kern, F.G. (1994). Emergence of MCF-7 cells overexpressing a transfected epidermal growth factor receptor (EGFR) under estrogen-depleted conditions: evidence for a role of EGFR in breast cancer growth and progression. *Cell Growth Differ,* 5, 1263-1274.

Murphy, L.C., Simon, S.L., Parkes, A., Leygue, E., Dotzlaw, H., Snell, L., Troup, S., Adeyinka, A. and Watson, P.H. (2000). Altered expression of estrogen receptor coregulators during human breast tumorigenesis. *Cancer Res*, 60, 6266-6271.

Myers, E., Fleming, F.J., Crotty, T.B., Kelly, G., McDermott, E.W., O'higgins, N.J., Hill, A.D. and Young, L.S. (2004). Inverse relationship between ER-beta and SRC-1 predicts outcome in endocrine-resistant breast cancer. *Br. J. Cancer*, 91, 1687-1693.

Myers, E., Hill, A.D.K., Kelly, G., McDermott, E.W., O'Higgins, N.J., Buggy, Y. and Young, L.S. (2005). Associations and interactions between Ets-1 and Ets-2 and coregulatory proteins, SRC-1, AIB1, and NCoR in breast cancer. *Clin Cancer Res.*, 11, 2111-2122.

Na, S.Y., Lee, S.K., Han, S.J., Choi, H.S., Im, S.Y. and Lee, J.W. (1998). Steroid receptor coactivator-1 interacts with the p50 subunit and coactivates nuclear factor kappaB-mediated transactivations. *J. Biol. Chem.*, 273, 10831-10834.

Newman, S.P., Bates, N.P., Vernimmen, D., Parker, M.G. and Hurst HC. (2000). Cofactor competition between the ligand-bound oestrogen receptor and an intron 1 enhancer leads to oestrogen repression of ERBB2 expression in breast cancer. *Oncogene*, 19, 490-497.

Osborne, C., Elledge, R. and Fuqua S. (1996). Estrogen receptors in breast cancer therapy. *Sci. Am. Sci. Med.*, 3, 32–41.

Osborne, C.K., Bardou, V., Hopp, T.A., Chamness, G.C., Hilsenbeck, S.G., Fuqua, S.A., Wong, J., Allred, D.C., Clark, G.M. and Schiff, R. (2003). Role of the estrogen receptor coactivator AIB1 (SRC-3) and HER-2/neu in tamoxifen resistance in breast cancer. *J. Natl. Cancer Inst*, 95,353-361.

Pearce, S.T., Liu, H. and Jordan, V.C. (2003). Modulation of estrogen receptor a function and stability by tamoxifen and a critical amino acid (Asp-538) in helix 12. *J. Biol. Chem*, 278, 7630-7638.

Pietras, R.J., Arboleda, J., Reese, D.M., Wongvipat, N., Pegram, M.D., Ramos, L., Gorman, C.M., Parker, M.G., Sliwkowski, M.X. and Slamon, D.J. (1995). HER-2 tyrosine kinase pathway targets estrogen receptor and promotes hormone-independent growth in human breast cancer cells. *Oncogene*, 10, 2435-2446.

Planas-Silva, M.D., Shang, Y., Donaher, J.L., Brown, M. and Weinberg, R.A. (2001). AIB1 enhances estrogen-dependent induction of cyclin D1 expression. *Cancer Res*, 61, 3858-3862.

Qin, L., Liao, L., Redmond, A., Young, L., Yuan, Y., Chen, H., O'Malley, B.W. and Xu, J. (2008). The AIB1 oncogene promotes breast cancer metastasis by activation of PEA3-mediated matrix metalloproteinase 2 (MMP2) and MMP9 expression. *Mol. Cell Biol*, 28, 5937-5950.

Qin, L., Liu, Z., Chen, H. and Xu, J. (2009). The steroid receptor coactivator-1 regulates twist expression and promotes breast cancer metastasis. *Cancer Res*, 69, 3819-3827.

Read, L.D., Keith, D., Slamon, D.J. and Katzenellenbogen, B.S. (1990). Hormonal modulation of HER-2/neu protooncogene messenger ribonucleic acid and p185 protein expression in human breast cancer cell lines. *Cancer Res*, 50, 3947-3951.

Redmond, A.M., Bane, F.T., Stafford, A.T., McIlroy, M., Dillon, M.F., Crotty, T.B., Hill, A.D. and Young, L.S. (2009). Coassociation of estrogen receptor and p160 proteins predicts resistance to endocrine treatment; SRC-1 is an independent predictor of breast cancer recurrence. *Clin. Cancer Res*, 15, 2098-2106.

Romano, A., Adriaens, M., Kuenen, S., Delvoux, B., Dunselman, G., Evelo, C. and Groothuis, P. (2010). Identification of novel ER-alpha target genes in breast cancer cells: gene- and cell-selective co-regulator recruitment at target promoters determines the response to 17beta-estradiol and tamoxifen. *Mol. Cell Endocrinol*, 314, 90-100.

Rowan, B.G., Garrison, N., Weigel, N.L. and O'Malley, B.W. (2000). 8-bromo-cyclic AMP induces phosphorylation of two sites in SRC-1 that facilitate ligand-independent activation of the chicken progesterone receptor and are critical for functional cooperation between SRC-1 and CREB binding protein. *Mol. Cell Biol,* 20, 8720-8730.

Sabbah, M., Courilleau, D., Mester, J. and Redeuilh, G. (1999). Estrogen induction of the cyclin D1 promoter: involvement of a cAMP response-like element. *Proc. Natl. Acad. Sci. USA*, 96, 11217–11222.

Santen, R.J., Song, R.X., Zhang, Z., Kumar, R., Jeng, M.H., Masamura, A., Lawrence, J. Jr., Berstein, L. and Yue, W. (2005). Long-term estradiol deprivation in breast cancer cells up-regulates growth factor signaling and enhances estrogen sensitivity.*Endocr. Relat. Cancer*, 12 Suppl 1, S61-73.

Shah, Y.M. and Rowan, B.G. (2005). The Src kinase pathway promotes tamoxifen agonist action in Ishikawa endometrial cells through phosphorylation-dependent stabilization of estrogen receptor a promoter interaction and elevated steroid receptor coactivator 1 activity. *Mol. Endocrinol*, 19, 732-748.

Shang, Y. and Brown, M. (2002). Molecular determinants for the tissue specificity of SERMs. *Science*, 295, 2465-2468.

Shiau, A.K., Barstad, D., Loria, P.M., Cheng, L., Kushner, P.J., Agard, D.A. and Greene, G.L. (1998). The structural basis of estrogen receptor/coactivator recognition and the antagonism of this interaction by tamoxifen. *Cell*, 95, 927-937.

Shiau, A.K., Barstad, D., Radek, J.T., Meyers, M.J., Nettles, K.W., Katzenellenbogen, B.S., Katzenellenbogen, J.A., Agard, D.A and Greene, G.L. (2002). Structural characterization of a subtype-selective ligand reveals a novel mode of estrogen receptor antagonism. *Nat. Struct. Biol*, 9, 359-364.

Shin, I., Miller, T. and Arteaga, C.L. (2006). ErbB receptor signaling and therapeutic resistance to aromatase inhibitors. *Clin. Cancer Res*, 12, 1008s-1012s.

Shou, J., Massarweh, S., Osborne, C.K., Wakeling, A.E., Ali, S., Weiss, H. and Schiff, R. (2004). Mechanisms of tamoxifen resistance: Increased estrogen receptor-HER2/neu cross-talk in ER/HER2-positive breast cancer. *J. Nat. Cancer Inst*, 96, 926-935.

Slamon, D.J., Clark, G.M., Wong, S.G., Levin, W.J., Ullrich, A. and McGuire, W.L. (1987). Human breast cancer: correlation of relapse and survival with amplification of the HER-2/neu oncogene. *Science*, 235, 177–182.

Smith, C.L., Nawaz, Z. and O'Malley, B.W. (1997). Coactivator and corepressor regulation of the agonist/antagonist activity of the mixed antiestrogen, 4-hydroxytamoxifen. *Mol. Endocrinol*, 11, 657-666.

Stearns, V., Johnson, M.D., Rae, J.M., Morocho, A., Novielli, A., Bhargava, P., Hayes, D.F., Desta, Z., and Flockhart, D.A. (2003). Active tamoxifen metabolite plasma concentrations after coadministration of tamoxifen and the selective serotonin reuptake inhibitor paroxetine. *J. Natl. Cancer Inst*, 95, 1758–1764.

Sun, M., Paciga, J.E., Feldman, R.I., Yuan, Z., Coppola, D., Lu, Y.Y., Shelley, S.A., Nicosia, S.V. and Cheng, J.Q. (2001). Phosphatidylinositol-3-OH kinase (PI3K)/AKT2, activated in breast cancer, regulates and is induced by estrogen receptor a (ERa) via interaction between ERa and PI3K. *Cancer Res*, 61, 5985-5991.

Torchia, J., Rose, D.W., Inostroza, J., Kamei, Y., Westin, S., Glass, C.K. and Rosenfeld, M.G. (1997). The transcriptional coactivator p/CIP binds CBP and mediates nuclear-receptor function. *Nature*, 387, 677-684.

Torres-Arzayus, M.I., Font de Mora, J., Yuan, J., Vazquez, F., Bronson, R., Rue, M., Sellers, W.R. and Brown, M. (2004). High tumor incidence and activation of the PI3K/AKT pathway in transgenic mice define AIB1 as an oncogene. *Cancer Cell*, 6, 263-274.

Torres-Arzayus, M.I., Zhao, J., Bronson, R. and Brown, M. (2010). Estrogen-dependent and estrogen-independent mechanisms contribute to AIB1-mediated tumor formation. *Cancer Res*, 70, 4102-4111.

Umayahara, Y., Kawamori, R., Watada, H., Imano, E., Iwama, N., Morishima, T., Yamasaki, Y., Kajimoto, Y.and Kamada, T. (1994). Estrogen regulation of the insulin-like growth factor I gene transcription involves an AP-1 enhancer. *J. Biol. Chem*, 269, 16433-16442.

Wang, L.H., Yang, X.Y., Zhang, X., An, P., Kim, H.J., Huang, J., Clarke, R., Osborne, C.K., Inman, J.K., Appella, E. and Farrar, W.L. (2006). Disruption of estrogen receptor DNA-binding domain and related intramolecular communication restores tamoxifen sensitivity in resistant breast cancer. *Cancer Cell*, 10, 487-499.

Wang, S.,Yuan, Y., Liao, L., Kuang, S.Q., Tien, J.C., O'Malley, B.W. and Xu, J. (2009). Disruption of the SRC-1 gene in mice suppresses breast cancer metastasis without affecting primary tumor formation. *Proc. Natl. Acad. Sci. USA*, 106, 151-156.

Warri, A.M., Laine, A.M., Majasuo, K.E., Alitalo, K.K. and Harkonen, P.L. (1991). Estrogen suppression of erbB2 expression is associated with increased growth rate of ZR-75-1 human breast cancer cells in vitro and in nude mice. *Int. J. Cancer*, 49, 616-623.

Webb, P., Nguyen, P. and Kushner, P.J. (2003). Differential SERM effects on corepressor binding dictate ER a activity in vivo. *J. Biol. Chem*, 278, 6912-6920.

Webb, P., Nguyen, P., Shinsako, J., Anderson, C., Feng, W., Nguyen, M.P., Chen, D., Huang, S.M., Subramanian, S., McKinerney, E., Katzenellenbogen, B.S., Stallcup, M.R. and Kushner, P.J. (1998). Estrogen receptor activation function 1 works by binding p160 coactivator proteins. *Mol. Endocrinol*, 12, 1605-1618.

Weisz, A. and Rosales, R. (1990). Identification of an estrogen response element upstream of the human c-fos gene that binds the estrogen receptor and the AP-1 transcription factor. *Nucleic Acids Res*, 18, 5097–5106.

Wu, R.C., Qin, J., Hashimoto, Y., Wong, J., Xu, J., Tsai, S.Y., Tsai, M.J. and O'Malley, B.W. (2002). Regulation of SRC-3 (pCIP/ACTR/AIB-1/RAC-3/TRAM-1) coactivator activity by I kappa B kinase. *Mol. Cell Biol*, 22, 3549-3561.

Wu, R.C., Qin, J., Yi, P., Wong, J., Tsai, S.Y., Tsai, M.J. and O'Malley, B.W. (2004). Selective phosphorylation of the SRC-3/AIB1 coactivator integrate genomic responses to multiple cellular signaling pathways. *Mol. Cell*, 15, 937-949.

Xu, J. and Li, Q. (2003). Review of the in vivo functions of the p160 steroid receptor coactivator family. *Mol. Endo*, 17, 1681-92.

Xu, J., Liao, L., Ning, G., Yoshida-Komiya, H., Deng, C. and O'Malley, B.W. (2000). The steroid receptor SRC-3 (p/CIP/RAC3/AIB1/ACTR/TRAM-1) is required for normal growth, puberty, female reproductive function and mammary gland development. *Proc. Natl. Acad. Sci. USA*, 97, 6379-6384.

Xu, J. and O'Malley, B.W. (2002). Molecular mechanisms and cellular biology of the steroid receptor coactivator (SRC) family in steroid receptor function. *Rev. Endocr. Metab. Disord*, 3, 185-192.

Xu, J., Wu, R.C. and O'Malley, B.W. (2009). Normal and cancer-related functions of the p160 steroid receptor co-activator (SRC) family. *Nat. Rev. Cancer*, 9, 615-630.

Yager, J.D. and Davidson, N.E. (2006). Estrogen carcinogenesis in breast cancer. *N. Engl. J. Med*, 354, 270-282.

Yan, J., Tsai, S.Y. and Tsai, M.J. (2006). SRC-3/AIB1: transcriptional coactivator in oncogenesis. *Acta Pharmacol. Sin*, 27, 387-394.

Yan, J., Erdem, H., Li, R., Cai, Y., Ayala, Y., Ittmann, M., Yu-Lee, L.-Y., Tsai, S.Y. and Tsai, M.J. (2008). Steroid receptor coactivator-3/AIB1 promotes cell migration and invasiveness through focal adhesion turnover and matrix metalloproteinase expression. *Cancer Res*, 68, 5460-5467.

Zhao, C., Yasui, K., Lee, C.J., Kurioka, H., Hosokawa, Y., Oka, T. and Inazawa, J. (2003). Elevated expression levels of NCOA3, TOP1, and TFAP2C in breast tumors as predictors of poor prognosis. *Cancer*, 98,18-23.

Zhao, W., Zhang, Q., Kang, X., Jin, S. and Lou, C. (2009). AIB1 is required for the acquisition of epithelial growth factor receptor-mediated tamoxifen resistance in breast cancer cells. *Biochem. Biophys. Res. Commun*, 380, 699–704.

Zhou, G., Hashimoto, Y., Kwak, I., Tsai, S.Y. and Tsai, M.J. (2003). Role of the steroid receptor coactivator SRC-3 in cell growth. *Mol. Cell Biol*, 23, 7742-7755.

In: Nuclear Receptors
Editors: M. K. Bates and R. M. Kerr

ISBN: 978-1-61209-980-4

Chapter 3

THE FUNCTIONAL ROLE OF THE GLUCOCORTICOID RECEPTOR (GR) AND NUR77 IN THYMOCYTE DEVELOPMENT

***Noriko Tosa*[1], *Takahiro Fukumoto*[2] *and Tadaaki Miyazaki*[3,*]**

[1]Institute for Animal Experimentation, Hokkaido University Graduate School of Medicine, Kita-15, Nishi-7, Kita-ku, Sapporo, Hokkaido, 060-8638, Japan
[2]Research Center for Infection-associated Cancer, Institute for Genetic Medicine, Hokkaido University, Kita-15, Nishi-7, Kita-ku, Sapporo, 060-0815, Japan
[3]Department of Bioresources, Hokkaido University Research Center for Zoonosis Control, North 20, West 10, Kita-ku, Sapporo, 001-0020, Japan

ABSTRACT

Nuclear receptors are involved in various aspects of intracellular signal transduction on a range of tissue and play an important role as regulators in numerous essential biological functions. In the thymus, these nuclear receptors also participate in positive or negative selection during T cell development.

In particular, the glucocorticoid receptor (Gr) and Nur77 play central role s in apoptosis induction mediated by the T cell receptor (TCR) in mature thymocytes or glucocorticoids (GCs) in immature thymocytes,

[*]E-mail: miyazaki@czc.hokudai.ac.jp, Tel. and Fax: +81-11-706-7314

respectively. Recently, we demonstrated that death-associated protein 3 (DAP3) was critical for TCR-mediated induction of apoptosis downstream of Nur77 in immature thymocytes. The DAP3 is an evolutionarily conserved GTP binding protein that plays a number of roles in normal mitochondrial physiology and in apoptosis induced via the tumor necrosis factor (TNF) family of death receptors. This chapter reviews recent studies of the signal transduction mediated by Gr and Nur77 in thymocyte development, focusing on signaling molecules, such as DAP3, involved in the signaling pathways of Gr or Nur77. Briefly, discussion which have attracted attention are summarized as follows: 1) signaling molecules interacting with Gr or Nur77, 2) the functional role of Gr or Nur77 in subcellular localization, 3) the function of genes subject to induction by Gr or Nur77, 4) crosstalk and its physiological importance in the signal transduction mediated by Gr and Nur77.

INTRODUCTION

During T cell development in the thymus, precursor thymocytes are guided through numerous checkpoints to ensure the generation of sufficient numbers of T cells bearing functional antigen receptors that do not cross-react with self-antigens. An important checkpoint in T cell development occurs at the $CD4^{+}CD8^{+}$ double positive (DP) stage where interaction of the T cell receptors (TCR) with major histocompatibility complex (MHC)/self-peptides on thymic epithelial and stromal cells initiates selection processes that lead to death by neglect, positive, or negative selection (Starr *et al.* 2003; Bommhardt *et al.* 2004). At this stage, the majority of DP thymocytes die due to their inefficiency in interacting with self-ligands in the context of MHC, death by neglect. It is believed that glucocorticoids (GCs) are in part responsible for cell death in the case of death by neglect. According to the avidity model of selection, DP cells with TCRs that interact with moderate affinity to self-ligands will be positively selected and further differentiate into mature functional T cells (Bevan 1997). High-affinity interaction with self-MHC, which would be equivalent to autoreactivity in the periphery, puts DP cells on the path to apoptosis and thus results in negative selection.

A number of nuclear receptors participate in each selection process during T cell development in the thymus (Starr *et al.* 2003; Bommhardt *et al.* 2004). In particular, Nur77 and glucocorticoid receptor (Gr) play pivotal roles in apoptosis induction mediated by TCR in immature thymocytes or in GC-induced apoptosis in immature thymocytes (He 2002; Herold *et al.* 2006). The Gr belongs to the nuclear receptor superfamily which is characterized by its commonality in the arrangement of functionally distinct domains mediating transactivation, DNA binding, nuclear localization, dimerization, and ligand recognition (Mangelsdorf *et al.* 1995). GCs such as cortisol in humans and corticosterone in rodents, a class of steroid hormones, diffuse passively into cells, where they bind to the Gr, and display potent immunomodulatory activities including the ability to induce T lymphocyte apoptosis (Beato *et al.* 1995; Herold *et al.* 2006). This GC-induced apoptosis was one of the first recognized forms of programmed cell death, however it remains poorly understood (Wyllie 1980). The nuclear receptor Nur77 is an orphan nuclear receptor belonging to the steroid/thyroid hormone receptor superfamily, and is the transcription factor responsible for inducing apoptosis (Moll *et al.* 2006). Further, Nur77 is activated by a variety of agents stimulating apoptosis induction, and among these TCR stimulation is known to be a potent activator of Nur77 transcription (Liu *et al.* 1994; Woronicz *et al.* 1994). A previous report has indicated that thymocytes from transgenic mice that express a dominant-negative form of Nur77 (DN-Nur77) are protected from negative selection, and conversely, thymocytes from transgenic mice that express wild-type Nur77 exhibit promoted negative selection (Calnan *et al.* 1995). Despite a large body of evidence supporting the paradigm that Nur77-mediated apoptosis induction is critical for negative selection, the signaling pathway downstream of Nur77 in TCR-mediated apoptosis in immature thymocytes is not fully understood and there is no agreement as to how Nur77 induces apoptosis.

Recently, we demonstrated that death-associated protein 3 (DAP3) was critical for TCR-mediated induction of apoptosis downstream of Nur77 in immature thymocytes (Tosa *et al.* 2010). The DAP3 is an evolutionarily conserved GTP binding protein that plays a number of roles in normal mitochondrial physiology and in apoptosis induced via tumor necrosis factor (TNF) family death receptors (Kim *et al.* 2007). Also, DAP3 expresses ubiquitously in various kinds of tissue including the immune system, such as in the thymus, of humans and mice (Berger *et al.* 2000; Hirota *et al.* 2004). This DAP3 is a GTP-binding protein that has been identified as a positive mediator in interferon (IFN)-γ-induced cell death (Kissil *et al.* 1995). The *DAP3* gene codes for a 46 kDa protein with a potential P-loop motif, a potential nuclear receptor-interacting domain, and a putative cleavage site for N-terminal mitochondrial important sequences (Kissil *et al.* 1995; Hulkko *et al.* 2000; Morgan *et al.* 2001). A previous study has shown that DAP3 binds a number of nuclear receptors including Gr (Hulkko *et al.* 2000). We have been revealed the physiological function of DAP3 and have etc. We have demonstrated that DAP3 in the cytosol functions as a pro-apoptotic adaptor molecule between TRAIL-receptors and FAS-associated death domain protein (FADD) (Miyazaki and Reed 2001). It is reported that mitochondrial DAP3 regulates cellular senescence though oxidative stress response (Murata *et al.* 2006). Further, DAP3 is phosphorylated in an Akt-dependent manner, correlating with the suppression of DAP3-facilitated apoptosis in anoikis (Miyazaki *et al.* 2004), and interferon-β promoter stimulator 1 (IPS-1) binds DAP3, resulting in the induction of anoikis by caspase activation (Li *et al.* 2009). Recently, we identified a novel DAP3-binding protein termed death ligand signal enhancer (DELE) and found that DELE actually binds to DAP3 subsequent to the involvement in the apoptosis signal mediated by the death receptor (Harada *et al.* 2010).

In this chapter, we review recent studies of the signal transduction mediated by Gr and Nur77 in thymocyte development, focusing on signaling molecules such as DAP3, involved in the signaling pathways of Gr or Nur77. Briefly, the main points of the discussion is summarized as follows: 1) signaling molecules interacting with Gr or Nur77, 2) the functional role of Gr or Nur77 in subcellular localization, 3) the function of the genes subject to induction by Gr or Nur77, 4) crosstalk and the physiological importance of Gr and Nur77 in the signal transduction mediated by Gr and Nur77.

The Signaling Molecules Interacting with Gr or Nur77

Glucocorticoid receptor (Gr)

In the absence of hormones, Gr is found within a multimeric complex of heat shoch proteins 90 (Hsp90) and immunophilins in the cytoplasm of the thymus (Pratt *et al.* 1996)(Figure 1). Following dissociation from the Hsp90 and combination with GCs, Gr translocates into the nucleus. In the nucleus, Gr modulates transcription either by binding to DNA or via interaction with other transcription factors. Here Gr recognizes imperfect palindromic sequences, so-called glucocorticoid response elements (GREs), present in the promoter and enhancer regions of a variety of genes (Luisi *et al.* 1991) and Gr can drive the transcription from these response elements by using its surfaces as platforms for the docking of transcriptional coactivators that are capable of altering the local chromatin or recruiting and stabilizing the transcription machinery (Jenkins *et al.* 2001). Different from these DNA binding-dependent activities, Gr also regulates gene expression by interfering with other transcription factors such as NF-κB (Heck *et al.* 1997), AP-1 (Heck *et al.* 1994), NF-AT (Vacca *et al.* 1992), CREB (Imai *et al.* 1993) and Stat5 (Stoecklin *et al.* 1997). It is noteworthy that Gr can control gene expression in this way, mostly in a transrepressing manner, without binding to DNA itself. This was most convincingly demonstrated by introducing a point mutation into Gr which prevents it feom both dimerizing and binding DNA (Reichardt *et al.* 1998). When present in the germline of knock-in mice (GRdim), the mutated Gr was unable to mediate transcription from GREs but its ability to regulate transcription from AP-1-, NF-κB- and Stat5-driven promoters was not compromised (Tuckermann *et al.* 1999; Reichardt *et al.* 2001a; Reichardt *et al.* 2001b; Tronche *et al.* 2004).

Multiple signaling pathways appear to act on the initiation of GC-induced apoptosis (Lepine *et al.* 2005). Members of the BCL-2 family play an important role: for example, the thymocytes of mice null for both the Bax and Bak genes are resistant to GC-induced apoptosis (Rathmell *et al.* 2002). GC-induced apoptosis in thymocytes appears to act via BAX/BAK-mediated release of cytochrome c from mitochondria and activation of Caspase-9 (Wei *et al.* 2001). BCL-2 overexpressing T cell hybridoma is resistant to GC-induced apoptosis (Memon *et al.* 1995). The expression of Bim, a BH3-domain-only member of the BCL-2 family, has been shown to be induced by GCs (Wang *et al.* 2003a). Thymocytes from Bim-/- mice show significantly reduced sensitivity to GC-induced apoptosis (Bouillet *et al.* 1999) and downregulation of Bim expression in a B-cell line by RNA interference (RNAi) greatly reduces sensitivity to GCs (Abrams *et al.* 2004). While these signaling pathways appear to act on the initiation of GC-induced apoptosis, the upstream events are unclear.

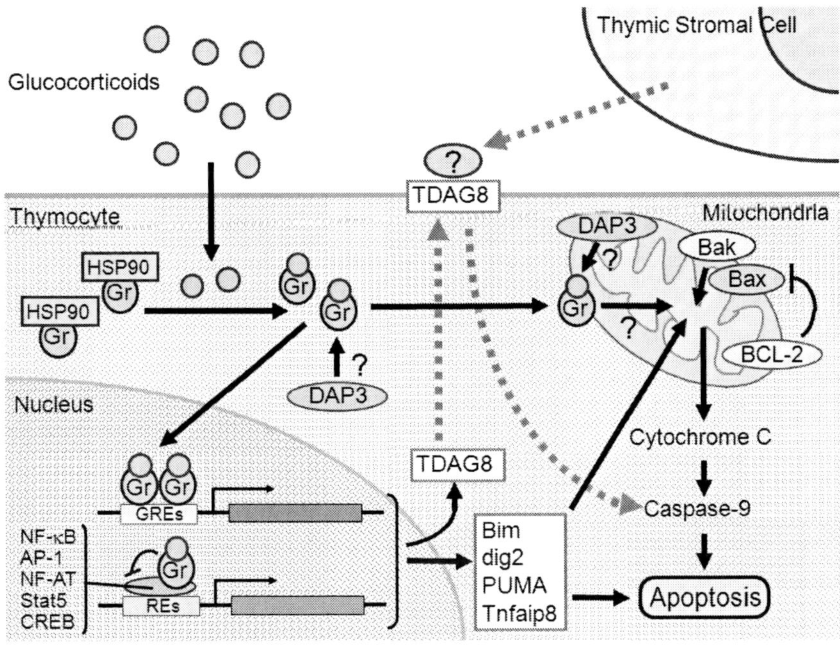

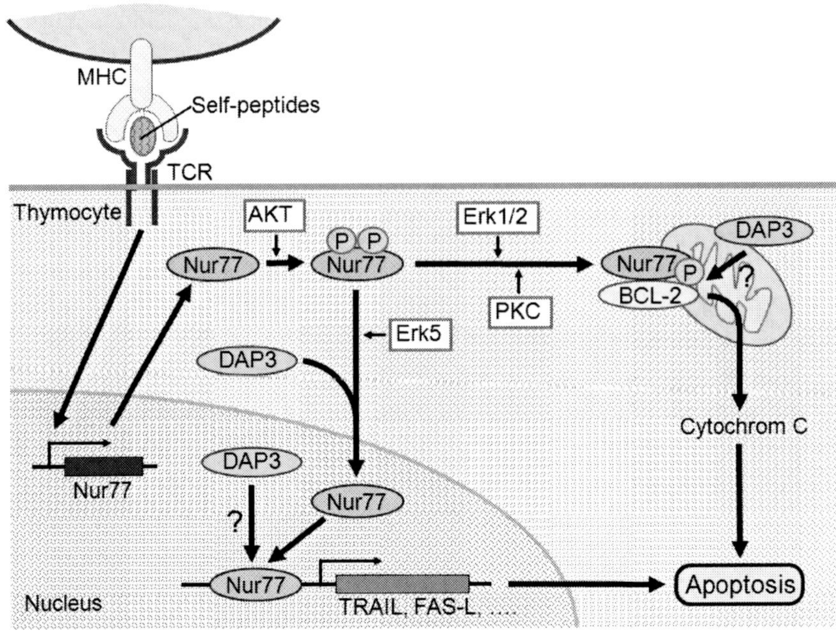

Nuclear receptor Nur77

Nur77, the orphan steroid receptor, is a highly phosphorylated protein present in T cells (Woronicz *et al.* 1995) and a number of protein kinase regulate the phosphorylation of Nur77 (Figure 2). The transcriptional activation of Nur77 is positively regulated by ERK5-mediated phosphorylation, while Nur77 is negatively regulated by Akt-mediated phosphorylation in TCR-mediated signal apoptosis. Fujii et al. reported that ERK5-mediated phosphorylation was indispensable for the positive regulation of the Nur77 function, as the inhibition of the ERK5 pathway resulted in the blockade of TCR-mediated apoptosis (Fujii *et al.* 2008). It has also been shown that Akt-mediated phosphorylation of Nur77, which phosphorylates Ser-350 in the DNA binding domain of Nur77, inhibits its DNA binding activity, resulting in a suppression of TCR-mediated signal apoptosis (Masuyama *et al.* 2001; Pekarsky *et al.* 2001). In sum, Nur77 is phosphorylated by Akt in cytoplasm and dephosphorylated in the nucleus. These results are consistent with our results that hyper-phosphorylated Nur77 (80 kDa) are predominantly localized in the cytsol fraction and that hypo-phosphorylated Nur77 (70 kDa) are localized in the nucleus fraction (Tosa *et al.* 2010). Recently, a kinase downstream of the ERK1/2 pathway has been shown to be able to effect the phosphorylation of Nru77 required for mitochondria translocation in DO11.10 (Wang *et al.* 2009). It is reported that Protein kinase C (PKC) also regulates mitochondria translocation of Nur77 and BCL-2 conformation changes during TCR-induced thymocyte apoptosis (Thompson *et al.* 2010). Putting all of the above together, these results suggest that the subcellular translocation of Nur77 in TCR-mediated apoptosis induction is determined by differences in kinas, that is, Nur77 phosphorylated by ERK1/2 or PKC is translocated into mitochondria and Nur77 phosphorylated by ERK5 or AKT is translocated into mitochondria.

It is considered that Nur77-mediated apoptosis induction in thymocytes activates two signal transductions, one with a mitochondria-dependent pathway and the other with a Nur77 transcriptional-dependent pathway. The Nur77 in mitochondria associates with BCL-2 and exposes its proapoptotic BH3 domain, and subsequently leads to apoptosis in the negative selection (Thompson and Winoto 2008). The Nur77 in the nucleus is activated without any known ligands and regulates the gene expression of proapoptotic protein such as Fas ligand (FasL), tumor necrosis factor (TNF)-related apoptosis-inducing ligand (TLAIL), and Nur77 downstream gene 1 (NDG1) (Rajpal *et al.* 2003). It has been shown that the C-terminal domain of Nur77 has the transactivation activity of the gene expression (Paulsen *et al.* 1992), however, the manner in which Nur77 regulates the gene expression of FasL, TLAIL, and NDG1 is not established. Related to this, we have demonstrated that DAP3 is located downstream of Nur77 in TCR-mediated signal apoptosis, as the translocation of DAP3 to the nucleus was blocked by overexpression of DN-Nur77, which blocked TCR-mediated apoptosis induction in DO11.10 T cell hybridomas after anti-CD3 antibody stimulation (Tosa *et al.* 2010). It is known that DAP3 has a NR box-like structure close to the N-terminal side of the P-loop motif (Hulkko *et al.* 2000). It is also known that DAP3 is phosphorylated in an Akt-dependent manner, correlating with the suppression of DAP3-facilitated apoptosis in anoikis (Miyazaki *et al.* 2004). Therefore, DAP3 may possibly translocate into the nucleus by binding to Nur77 directly or by phosphorylation in an Akt-dependent manner.

The Functional Role of Gr or Nur77 in Subcellular Localization

Gr

It has been considered that GCs first activate the Gr transcriptional-dependent pathway and subsequently the mitochondria-dependent pathway in GC-induced apoptosis induction of thymocytes. However, several studies have recently reported that Gr directly translocates to mitochondria in GC-induced apoptosis induction. Sionov *et al.* (2006) reported that mitochondrial and nuclear translocation of Gr were differently regulated, and that mitochondrial Gr translocation correlates with susceptibility to GC-induced apoptosis. Talaber *et al.* (2009) described that mitochondrial translocation of Gr in DP thymocytes correlates with the sensitivity to GC-induced apoptosis. These results suggest that this novel mechanism is a possible alternative Gr-signaling pathway, which could mediate the rapid apoptotic response in thymocytes, however, details of the process leading to the translocated GC-Gr complex initiation of the apoptotic cascade in mitochondria are still to be elucidated. It is also well established that the proteins targeting mitochondria associate with chaperones that play a role in the mitochondrial translocation. One of the commonly encountered chaperones in this category is Hsp70 (Dittmar *et al.* 1997; Cvoro *et al.* 1998). Previous studies have shown that Gr forms protein complexes with Hsp70/90 and BCL-2-associated athanogene (Bag-1) in response to GCs treatment (Cato and Mink 2001; Filipovic *et al.* 2005; Pratt *et al.* 2006). Here it may be noteworthy that DAP3 also binds to Gr as well as Hsp90 directly (Hulkko *et al.* 2000; Hulkko and Zilliacus 2002). However the mechanism of the translocation of GC-Gr complexes in thymocyte into mitochondria is not established.

Nur77

It has been considered that the subcellular localization of Nur77 is tanslocated from cytosol to the nucleus after TCR stimulation in thymocytes. However, several studies recently reported that Nur77 was localized in the mitochondria fraction as well as in the cytosol and nuclear fractions in thymocytes or DO11.10 cells after treatment with anti-CD3 antibody or PMA plus ionomycin (Stasik *et al.* 2007; Thompson and Winoto 2008; Wang *et al.* 2009). It has also been found that Nur77 was localized in the mitochondria fraction as well as in the cytosol fraction and translocated to the nuclear fraction in DO11.10 cells after treatment with anti-CD3 antibody (Tosa *et al.* 2010). Here, there are differences in the way of translocation to the mitochondria fraction. The three former studies report that Nur77 was detected in mitochondria fraction after appearance in nuclear fraction (Stasik *et al.* 2007; Thompson and Winoto 2008; Wang *et al.* 2009). Our results demonstrate that Nur77 was present in the nuclear fraction after the appearance in the mitochondria fraction (Tosa *et al.* 2010). These results suggest the possibility of at least two pathways for the translocation of Nur77 into the mitochondria by TCR stimulation.

How Nur77 regulates TCR-mediated apoptosis induction remains to be established. Some studies suggest that its transcriptional activity is critical, while others suggest that its interactions with mitochondrial proteins are responsible for its apoptotic activity. Stasik *et al.* (2007) reported that ionomycin induces translocation of Nur77 to the nucleus as well as to mitochondria. Thompson and

Winoto (2008) demonstrated that during the early phase of thymocyte apoptosis, Nur77 and another member of the Nur77 family, Nor-1, translocate from the nucleus to the mitocondria where they bind BCL-2 and their association with BCL-2 exposes the proapoptotic BH3 domain within BCL-2. Therefore, it is conceivable that Nur77 translocated into mitochondria interact with BCL-2 and that their association with BCL-2 expresses the BH3 domain within BCL-2, resulting in induction of cytochrom c release during TCR-induced thymocyte apoptosis. It is also conceivable that Nur77 in the nucleus is activated without any known ligands and regulates the gene expression of proapoptotic protein such as FasL, TLAIL, and NDG1 (Rajpal *et al.* 2003), however, the manner in which Nur77 regulates the gene expression is not established. We have demonstrated that Nur77 is crucial for nuclear translocation of DAP3 in the apoptosis induced by TCR stimulation. Although the data presented here suggest that nuclear translocation of DAP3 is important for TCR induced apoptosis, the function of DAP3 molecular in the nucleus in apoptosis induction is still not clearly established. It is known that DAP3 has a NR box-like structure close to the N-terminal side of the P-loop motif (Hulkko *et al.* 2000). Recently, it was reported that the Fas apoptosis inhibitory molecule regulates TCR induced apoptosis of thymocytes by modulating Akt activation and Nur77 expression (Huo *et al.* 2010). This suggests the possibility that DAP3 may bind to Nur77 directly, and subsequently regulate the transcription of apoptotic genes, or possibly affect the transcriptional regulation of Nur77 by binding to an unknown transcription factor.

The Function of the Genes Subject to Induction by Gr or Nur77

Gr

In recent years a number of approaches to identify genes subject to GCs induction have been proposed, but only few convincing candidates were identified, TDAG8 (Tosa et al. 2003), dig2 (Wang et al. 2003b), Bim (Bouillet et al. 2002; Wang et al. 2003a), PUMA (Han et al. 2001), Tnfaip8 (Woodward et al. 2010). It is considered that Bim and PUMA are the most promising candidates in the identified gene subjects to GCs induction during the initiation of GC-induced apoptosis, because GC-induced apoptosis in Bim- or PUMA-deficient mice was partially impaired compared to wild-type animals (Bouillet et al. 1999; Jenkins et al. 2001; Marsden et al. 2002; Jeffers et al. 2003; Villunger et al. 2003). Although the identified gene subjects to GCs induction are almost all present in cytosol, TDAG8 appears to be localized in the plasma membrane, as TDAG8 encodes a putative G protein-coupled receptor containing seven transmembrane segments

(Kyaw et al. 1998). We have previously demonstrated that TDAG8 expression was involved in GC-induced signals to activate Caspase-9, -8, and, -3 for subsequent apoptosis induction in DP thymocytes (Tosa et al. 2003). Recently, it has been shown that the special microenvironment of the thymus plays a central role in thymocyte development (Ladi et al. 2006; Nitta et al. 2008). The complex mesh of thymic stromal cells is composed of thymic epithelial cells, macrophages, dendritic cells, and fibroblasts, providing a vast surface for direct cell-cell interactions and producing soluble factors regulating thymocyte development, intrathymic migration and selection processes (Anderson and Jenkinson 2001; Takahama et al. 2008). Important molecules taking part in cell-cell interactions are adhesion molecules, chemokine receptors (Savino et al. 2004; Kurobe et al. 2006; Nitta et al. 2009), Wnt (Pongracz et al. 2003), and MHC (Takahama et al. 2010; Klein and Kyewski 2000). Ligands of TDAG8 are not identified here, however it is still possible that TDAG8 is affected by factors secreted from thymic stromal cells.

Nur77

The apoptotic activity of Nur77 has been demonstrated to correlate with its transcriptional activity in the nucleus (Kuang *et al.* 1999), suggesting that Nur77 functions to induce proapoptotic gene expression. Also, Nur77 transcriptionally regulates proapoptotic gene expression including FasL, TRAIL and NDG-1 (Rajpal *et al.* 2003).

However, signaling proteins downstream of FasL, TRAIL and NDG-1 like FADD and caspase-8 are not required for negative selection. A dominant-negative protein of the FADD, which can simultaneously inhibit signaling from multiple death receptors, showed no effect on the negative selection (Newton *et al.* 1998). The analysis of conditional FADD: GFP knockout mice reveals that FADD is not necessary in thymic development since its thymocytes were resistant to apoptosis induced by Fas, TNF, and TCR restimulation but the thymocyte development was normal (Zhang *et al.* 2005).

The role of the TLAIL in negative selection is not generally agreed on (Cretney *et al.* 2003; Lamhamedi-Cherradi *et al.* 2003; Corazza *et al.* 2004). Studies with gene-targeted mice show that a number of the caspases, caspase-1, -2, -3, -8, -9, and, -11, are not necessary in deletion of selfreactive thymoctes (Hara *et al.* 2002; Salmena *et al.* 2003). These results support data that there is an unknown pathway for apoptosis induction mediated by DAP3 independent of the FADD and caspase-8 dependent pathway in TCR induced apoptosis (Tosa *et al.* 2010), although recruitment of FADD followed by activation of caspase-8 is thought to be a major molecular mechanism in the DAP3-mediated induction of apoptosis (Miyazaki and Reed 2001). These results indicate that the expression of genes that have not been identified possibly get involved in the negative selection of thymocytes.

Crosstalk and Its Physiological Importance in the Signal Transduction Mediated by Gr and Nur77

Most models of positive and negative selection postulate that the strength of TCR signaling determines the fate of thymocytes. In the "mutual antagonism" model, TCR signaling involvement in selection is assigned to an interplay between GC- and TCR-induced apoptosis (Vacchio and Ashwell 2000). Signaling either through TCR or GR induces apoptosis, but when both receptors are simultaneously stimulated, the effects of the two are opposite (Iwata *et al.* 1991). Thus, thymocytes expressing a TCR with high affinity for the MHC/self-peptide undergoes negative selection, since the GR signaling is not sufficient to overcome the strong signal originating from TCR. For thymocytes with low-avidity TCRs undergo GC-induced cell death, since TCR signals cannot override the GR signaling in this case (Chung *et al.* 2002). In thymocytes with moderate-avidiy TCRs, the two signals neutralize each other and the affected cells do not become subject to apoptosis. There is a large body of studies about the "mutual antagonism" model, however, this model is not fully understood.

This chapter has discussed about mechanisms of Nur77- or Gr-mediated apoptosis induction in thymocytes. As previously stated, Nur77- or Gr-mediated apoptosis induction activates two common signal transductions, one with a transcriptional-dependent pathway and the other with a mitochondria-dependent pathway in the thymocytes.

Further, there are molecules such as BCL-2 and DAP3 commonly involved in both of the signal transductions mediated by Gr or Nur77; Bim is also involved in TCR-mediated apoptosis induction as well as in Gr-mediated apoptosis induction, although it is currently how Bim and Nur77 cooperate during the induction of negative selection. In Bim deficient mice there is a prominent inhibition of negative selection that is induced by injection of CD3ε-specific antibody, the superantigen SEB or an antigenic peptide (Bouillet *et al.* 2002). Palinkas *et al.* (2008) have reported that the two separate signaling pathways, TCR activation and Gr signals, merge in DP thymocytes at important apoptosis regulating points like BCL-2 and Gr, suggesting that a balanced interplay of the two is essential in DP cell survival. It is considered that these results support the "mutual antagonism" model. Further work and study of the crosstalk of TCR-mediated signaling pathways and Gr signaling pathways is required.

REFERENCES

Abrams, M. T., N. M. Robertson, K. Yoon and E. Wickstrom (2004) Inhibition of glucocorticoid-induced apoptosis by targeting the major splice variants of BIM mRNA with small interfering RNA and short hairpin RNA. *J. Biol. Chem.* 279, 55809-17.

Anderson, G. and E. J. Jenkinson (2001) Lymphostromal interactions in thymic development and function. *Nat. Rev. Immunol.* 1, 31-40.

Beato, M., P. Herrlich and G. Schutz (1995) Steroid hormone receptors: many actors in search of a plot. *Cell.* 83, 851-7.

Berger, T., M. Brigl, J. M. Herrmann, V. Vielhauer, B. Luckow, D. Schlondorff and M. Kretzler (2000) The apoptosis mediator mDAP-3 is a novel member of a conserved family of mitochondrial proteins. *J. Cell Sci.* 113 (Pt 20), 3603-12.

Bevan, M. J. (1997) In thymic selection, peptide diversity gives and takes away. *Immunity.* 7, 175-8.

Bommhardt, U., M. Beyer, T. Hunig and H. M. Reichardt (2004) Molecular and cellular mechanisms of T cell development. *Cell Mol. Life Sci.* 61, 263-80.

Bouillet, P., D. Metcalf, D. C. Huang, D. M. Tarlinton, T. W. Kay, F. Kontgen, J. M. Adams and A. Strasser (1999) Proapoptotic Bcl-2 relative Bim required for certain apoptotic responses, leukocyte homeostasis, and to preclude autoimmunity. *Science.* 286, 1735-8.

Bouillet, P., J. F. Purton, D. I. Godfrey, L. C. Zhang, L. Coultas, H. Puthalakath, M. Pellegrini, S. Cory, J. M. Adams and A. Strasser (2002) BH3-only Bcl-2 family member Bim is required for apoptosis of autoreactive thymocytes. *Nature.* 415, 922-6.

Calnan, B. J., S. Szychowski, F. K. Chan, D. Cado and A. Winoto (1995) A role for the orphan steroid receptor Nur77 in apoptosis accompanying antigen-induced negative selection. *Immunity.* 3, 273-82.

Cato, A. C. and S. Mink (2001) BAG-1 family of cochaperones in the modulation of nuclear receptor action. *J. Steroid Biochem. Mol. Biol.* 78, 379-88.

Chung, H., Y. I. Choi, M. G. Ko and R. H. Seong (2002) Rescuing developing thymocytes from death by neglect. *J. Biochem. Mol. Biol.* 35, 7-18.

Corazza, N., G. Brumatti, S. Jakob, A. Villunger and T. Brunner (2004) TRAIL and thymocyte apoptosis: not so deadly? *Cell Death Differ*. 11 Suppl 2, S213-5.

Cretney, E., A. P. Uldrich, S. P. Berzins, A. Strasser, D. I. Godfrey and M. J. Smyth (2003) Normal thymocyte negative selection in TRAIL-deficient mice. *J. Exp. Med*. 198, 491-6.

Cvoro, A., J. Dundjerski, D. Trajkovic and G. Matic (1998) Association of the rat liver glucocorticoid receptor with Hsp90 and Hsp70 upon whole body hyperthermic stress. *J. Steroid Biochem. Mol. Biol*. 67, 319-25.

Dittmar, K. D., D. R. Demady, L. F. Stancato, P. Krishna and W. B. Pratt (1997) Folding of the glucocorticoid receptor by the heat shock protein (hsp) 90-based chaperone machinery. The role of p23 is to stabilize receptor.hsp90 heterocomplexes formed by hsp90.p60.hsp70. *J. Biol. Chem*. 272, 21213-20.

Filipovic, D., L. Gavrilovic, S. Dronjak and M. B. Radojcic (2005) Brain glucocorticoid receptor and heat shock protein 70 levels in rats exposed to acute, chronic or combined stress. *Neuropsychobiology*. 51, 107-14.

Fujii, Y., S. Matsuda, G. Takayama and S. Koyasu (2008) ERK5 is involved in TCR-induced apoptosis through the modification of Nur77. *Genes Cells*. 13, 411-9.

Han, J., C. Flemington, A. B. Houghton, Z. Gu, G. P. Zambetti, R. J. Lutz, L. Zhu and T. Chittenden (2001) Expression of bbc3, a pro-apoptotic BH3-only gene, is regulated by diverse cell death and survival signals. *Proc. Natl. Acad. Sci. U S A*. 98, 11318-23.

Hara, H., A. Takeda, M. Takeuchi, A. C. Wakeham, A. Itie, M. Sasaki, T. W. Mak, A. Yoshimura, K. Nomoto and H. Yoshida (2002) The apoptotic protease-activating factor 1-mediated pathway of apoptosis is dispensable for negative selection of thymocytes. *J. Immunol*. 168, 2288-95.

Harada, T., A. Iwai and T. Miyazaki (2010) Identification of DELE, a novel DAP3-binding protein which is crucial for death receptor-mediated apoptosis induction. *Apoptosis*. 15, 1247-55.

He, Y. W. (2002) Orphan nuclear receptors in T lymphocyte development. *J. Leukoc. Biol*. 72, 440-6.

Heck, S., K. Bender, M. Kullmann, M. Gottlicher, P. Herrlich and A. C. Cato (1997) I kappaB alpha-independent downregulation of NF-kappaB activity by glucocorticoid receptor. *EMBO J*. 16, 4698-707.

Heck, S., M. Kullmann, A. Gast, H. Ponta, H. J. Rahmsdorf, P. Herrlich and A. C. Cato (1994) A distinct modulating domain in glucocorticoid receptor monomers in the repression of activity of the transcription factor AP-1. *EMBO J*. 13, 4087-95.

Herold, M. J., K. G. McPherson and H. M. Reichardt (2006) Glucocorticoids in T cell apoptosis and function. *Cell Mol. Life Sci.* 63, 60-72.

Hirota, T., K. Obara, A. Matsuda, M. Akahoshi, K. Nakashima, K. Hasegawa, N. Takahashi, M. Shimizu, H. Sekiguchi, M. Kokubo, S. Doi, H. Fujiwara, A. Miyatake, K. Fujita, T. Enomoto, F. Kishi, Y. Suzuki, H. Saito, Y. Nakamura, T. Shirakawa and M. Tamari (2004) Association between genetic variation in the gene for death-associated protein-3 (DAP3) and adult asthma. *J. Hum. Genet.* 49, 370-5.

Hulkko, S. M., H. Wakui and J. Zilliacus (2000) The pro-apoptotic protein death-associated protein 3 (DAP3) interacts with the glucocorticoid receptor and affects the receptor function. *Biochem J.* 349 Pt 3, 885-93.

Hulkko, S. M. and J. Zilliacus (2002) Functional interaction between the pro-apoptotic DAP3 and the glucocorticoid receptor. *Biochem. Biophys. Res. Commun.* 295, 749-55.

Huo, J., S. Xu and K. P. Lam (2010) Fas apoptosis inhibitory molecule regulates T cell receptor-mediated apoptosis of thymocytes by modulating Akt activation and Nur77 expression. *J. Biol. Chem.* 285, 11827-35.

Imai, E., J. N. Miner, J. A. Mitchell, K. R. Yamamoto and D. K. Granner (1993) Glucocorticoid receptor-cAMP response element-binding protein interaction and the response of the phosphoenolpyruvate carboxykinase gene to glucocorticoids. *J. Biol. Chem.* 268, 5353-6.

Iwata, M., S. Hanaoka and K. Sato (1991) Rescue of thymocytes and T cell hybridomas from glucocorticoid-induced apoptosis by stimulation via the T cell receptor/CD3 complex: a possible in vitro model for positive selection of the T cell repertoire. *Eur. J. Immunol.* 21, 643-8.

Jeffers, J. R., E. Parganas, Y. Lee, C. Yang, J. Wang, J. Brennan, K. H. MacLean, J. Han, T. Chittenden, J. N. Ihle, P. J. McKinnon, J. L. Cleveland and G. P. Zambetti (2003) Puma is an essential mediator of p53-dependent and -independent apoptotic pathways. *Cancer Cell.* 4, 321-8.

Jenkins, B. D., C. B. Pullen and B. D. Darimont (2001) Novel glucocorticoid receptor coactivator effector mechanisms. *Trends Endocrinol. Metab.* 12, 122-6.

Kim, H. R., H. J. Chae, M. Thomas, T. Miyazaki, A. Monosov, E. Monosov, M. Krajewska, S. Krajewski and J. C. Reed (2007) Mammalian dap3 is an essential gene required for mitochondrial homeostasis in vivo and contributing to the extrinsic pathway for apoptosis. *FASEB J.* 21, 188-96.

Kissil, J. L., L. P. Deiss, M. Bayewitch, T. Raveh, G. Khaspekov and A. Kimchi (1995) Isolation of DAP3, a novel mediator of interferon-gamma-induced cell death. *J. Biol. Chem.* 270, 27932-6.

Klein, L. and B. Kyewski (2000) Self-antigen presentation by thymic stromal cells: a subtle division of labor. *Curr. Opin. Immunol.* 12, 179-86.

Kuang, A. A., D. Cado and A. Winoto (1999) Nur77 transcription activity correlates with its apoptotic function in vivo. *Eur. J. Immunol.* 29, 3722-8.

Kurobe, H., C. Liu, T. Ueno, F. Saito, I. Ohigashi, N. Seach, R. Arakaki, Y. Hayashi, T. Kitagawa, M. Lipp, R. L. Boyd and Y. Takahama (2006) CCR7-dependent cortex-to-medulla migration of positively selected thymocytes is essential for establishing central tolerance. *Immunity*. 24, 165-77.

Kyaw, H., Z. Zeng, K. Su, P. Fan, B. K. Shell, K. C. Carter and Y. Li (1998) Cloning, characterization, and mapping of human homolog of mouse T-cell death-associated gene. *DNA Cell Biol.* 17, 493-500.

Ladi, E., X. Yin, T. Chtanova and E. A. Robey (2006) Thymic microenvironments for T cell differentiation and selection. *Nat. Immunol.* 7, 338-43.

Lamhamedi-Cherradi, S. E., S. J. Zheng, K. A. Maguschak, J. Peschon and Y. H. Chen (2003) Defective thymocyte apoptosis and accelerated autoimmune diseases in TRAIL-/- mice. *Nat. Immunol.* 4, 255-60.

Lepine, S., J. C. Sulpice and F. Giraud (2005) Signaling pathways involved in glucocorticoid-induced apoptosis of thymocytes. *Crit. Rev. Immunol.* 25, 263-88.

Li, H. M., D. Fujikura, T. Harada, J. Uehara, T. Kawai, S. Akira, J. C. Reed, A. Iwai and T. Miyazaki (2009) IPS-1 is crucial for DAP3-mediated anoikis induction by caspase-8 activation. *Cell Death Differ.* 16, 1615-21.

Liu, Z. G., S. W. Smith, K. A. McLaughlin, L. M. Schwartz and B. A. Osborne (1994) Apoptotic signals delivered through the T-cell receptor of a T-cell hybrid require the immediate-early gene nur77. *Nature*. 367, 281-4.

Luisi, B. F., W. X. Xu, Z. Otwinowski, L. P. Freedman, K. R. Yamamoto and P. B. Sigler (1991) Crystallographic analysis of the interaction of the glucocorticoid receptor with DNA. *Nature*. 352, 497-505.

Mangelsdorf, D. J., C. Thummel, M. Beato, P. Herrlich, G. Schutz, K. Umesono, B. Blumberg, P. Kastner, M. Mark, P. Chambon and R. M. Evans (1995) The nuclear receptor superfamily: the second decade. *Cell.* 83, 835-9.

Marsden, V. S., L. O'Connor, L. A. O'Reilly, J. Silke, D. Metcalf, P. G. Ekert, D. C. Huang, F. Cecconi, K. Kuida, K. J. Tomaselli, S. Roy, D. W. Nicholson, D. L. Vaux, P. Bouillet, J. M. Adams and A. Strasser (2002) Apoptosis initiated by Bcl-2-regulated caspase activation independently of the cytochrome c/Apaf-1/caspase-9 apoptosome. *Nature*. 419, 634-7.

Masuyama, N., K. Oishi, Y. Mori, T. Ueno, Y. Takahama and Y. Gotoh (2001) Akt inhibits the orphan nuclear receptor Nur77 and T-cell apoptosis. *J. Biol. Chem.* 276, 32799-805.

Memon, S. A., M. B. Moreno, D. Petrak and C. M. Zacharchuk (1995) Bcl-2 blocks glucocorticoid- but not Fas- or activation-induced apoptosis in a T cell hybridoma. *J. Immunol.* 155, 4644-52.

Miyazaki, T. and J. C. Reed (2001) A GTP-binding adapter protein couples TRAIL receptors to apoptosis-inducing proteins. *Nat. Immunol.* 2, 493-500.

Miyazaki, T., M. Shen, D. Fujikura, N. Tosa, H. R. Kim, S. Kon, T. Uede and J. C. Reed (2004) Functional role of death-associated protein 3 (DAP3) in anoikis. *J. Biol. Chem.* 279, 44667-72.

Moll, U. M., N. Marchenko and X. K. Zhang (2006) p53 and Nur77/TR3 - transcription factors that directly target mitochondria for cell death induction. *Oncogene.* 25, 4725-43.

Morgan, C. J., C. Jacques, F. Savagner, Y. Tourmen, D. P. Mirebeau, Y. Malthiery and P. Reynier (2001) A conserved N-terminal sequence targets human DAP3 to mitochondria. *Biochem. Biophys. Res. Commun.* 280, 177-81.

Murata, Y., T. Wakoh, N. Uekawa, M. Sugimoto, A. Asai, T. Miyazaki and M. Maruyama (2006) Death-associated protein 3 regulates cellular senescence through oxidative stress response. *FEBS Lett.* 580, 6093-9.

Newton, K., A. W. Harris, M. L. Bath, K. G. Smith and A. Strasser (1998) A dominant interfering mutant of FADD/MORT1 enhances deletion of autoreactive thymocytes and inhibits proliferation of mature T lymphocytes. *EMBO J.* 17, 706-18.

Nitta, T., S. Murata, T. Ueno, K. Tanaka and Y. Takahama (2008) Thymic microenvironments for T-cell repertoire formation. *Adv. Immunol.* 99, 59-94.

Nitta, T., S. Nitta, Y. Lei, M. Lipp and Y. Takahama (2009) CCR7-mediated migration of developing thymocytes to the medulla is essential for negative selection to tissue-restricted antigens. *Proc. Natl. Acad. Sci. U S A.* 106, 17129-33.

Palinkas, L., G. Talaber, F. Boldizsar, D. Bartis, P. Nemeth and T. Berki (2008) Developmental shift in TcR-mediated rescue of thymocytes from glucocorticoid-induced apoptosis. *Immunobiology.* 213, 39-50.

Paulsen, R. E., C. A. Weaver, T. J. Fahrner and J. Milbrandt (1992) Domains regulating transcriptional activity of the inducible orphan receptor NGFI-B. *J. Biol. Chem.* 267, 16491-6.

Pekarsky, Y., C. Hallas, A. Palamarchuk, A. Koval, F. Bullrich, Y. Hirata, R. Bichi, J. Letofsky and C. M. Croce (2001) Akt phosphorylates and regulates the orphan nuclear receptor Nur77. *Proc. Natl. Acad. Sci. U S A*. 98, 3690-4.

Pongracz, J., K. Hare, B. Harman, G. Anderson and E. J. Jenkinson (2003) Thymic epithelial cells provide WNT signals to developing thymocytes. *Eur. J. Immunol.* 33, 1949-56.

Pratt, W. B., U. Gehring and D. O. Toft (1996) Molecular chaperoning of steroid hormone receptors. *EXS*. 77, 79-95.

Pratt, W. B., Y. Morishima, M. Murphy and M. Harrell (2006) Chaperoning of glucocorticoid receptors. *Handb. Exp. Pharmacol*, 111-38.

Rajpal, A., Y. A. Cho, B. Yelent, P. H. Koza-Taylor, D. Li, E. Chen, M. Whang, C. Kang, T. G. Turi and A. Winoto (2003) Transcriptional activation of known and novel apoptotic pathways by Nur77 orphan steroid receptor. *EMBO J.* 22, 6526-36.

Rathmell, J. C., T. Lindsten, W. X. Zong, R. M. Cinalli and C. B. Thompson (2002) Deficiency in Bak and Bax perturbs thymic selection and lymphoid homeostasis. *Nat. Immunol.* 3, 932-9.

Reichardt, H. M., K. Horsch, H. J. Grone, A. Kolbus, H. Beug, N. Hynes and G. Schutz (2001a) Mammary gland development and lactation are controlled by different glucocorticoid receptor activities. *Eur. J. Endocrinol.* 145, 519-27.

Reichardt, H. M., K. H. Kaestner, J. Tuckermann, O. Kretz, O. Wessely, R. Bock, P. Gass, W. Schmid, P. Herrlich, P. Angel and G. Schutz (1998) DNA binding of the glucocorticoid receptor is not essential for survival. *Cell.* 93, 531-41.

Reichardt, H. M., J. P. Tuckermann, M. Gottlicher, M. Vujic, F. Weih, P. Angel, P. Herrlich and G. Schutz (2001b) Repression of inflammatory responses in the absence of DNA binding by the glucocorticoid receptor. *EMBO J.* 20, 7168-73.

Salmena, L., B. Lemmers, A. Hakem, E. Matysiak-Zablocki, K. Murakami, P. Y. Au, D. M. Berry, L. Tamblyn, A. Shehabeldin, E. Migon, A. Wakeham, D. Bouchard, W. C. Yeh, J. C. McGlade, P. S. Ohashi and R. Hakem (2003) Essential role for caspase 8 in T-cell homeostasis and T-cell-mediated immunity. *Genes Dev.* 17, 883-95.

Savino, W., D. A. Mendes-Da-Cruz, S. Smaniotto, E. Silva-Monteiro and D. M. Villa-Verde (2004) Molecular mechanisms governing thymocyte migration: combined role of chemokines and extracellular matrix. *J. Leukoc. Biol.* 75, 951-61.

Sionov, R. V., O. Cohen, S. Kfir, Y. Zilberman and E. Yefenof (2006) Role of mitochondrial glucocorticoid receptor in glucocorticoid-induced apoptosis. *J. Exp. Med.* 203, 189-201.

Starr, T. K., S. C. Jameson and K. A. Hogquist (2003) Positive and negative selection of T cells. *Annu. Rev. Immunol.* 21, 139-76.

Stasik, I., A. Rapak, W. Kalas, E. Ziolo and L. Strzadala (2007) Ionomycin-induced apoptosis of thymocytes is independent of Nur77 NBRE or NurRE binding, but is accompanied by Nur77 mitochondrial targeting. *Biochim. Biophys. Acta.* 1773, 1483-90.

Stoecklin, E., M. Wissler, R. Moriggl and B. Groner (1997) Specific DNA binding of Stat5, but not of glucocorticoid receptor, is required for their functional cooperation in the regulation of gene transcription. *Mol. Cell Biol.* 17, 6708-16.

Takahama, Y., T. Nitta, A. Mat Ripen, S. Nitta, S. Murata and K. Tanaka (2010) Role of thymic cortex-specific self-peptides in positive selection of T cells. *Semin. Immunol.* 22, 287-93.

Takahama, Y., K. Tanaka and S. Murata (2008) Modest cortex and promiscuous medulla for thymic repertoire formation. *Trends Immunol.* 29, 251-5.

Talaber, G., F. Boldizsar, D. Bartis, L. Palinkas, M. Szabo, G. Berta, G. Setalo, Jr., P. Nemeth and T. Berki (2009) Mitochondrial translocation of the glucocorticoid receptor in double-positive thymocytes correlates with their sensitivity to glucocorticoid-induced apoptosis. *Int. Immunol.* 21, 1269-76.

Thompson, J., M. L. Burger, H. Whang and A. Winoto (2010) Protein kinase C regulates mitochondrial targeting of Nur77 and its family member Nor-1 in thymocytes undergoing apoptosis. *Eur. J. Immunol.* 40, 2041-9.

Thompson, J. and A. Winoto (2008) During negative selection, Nur77 family proteins translocate to mitochondria where they associate with Bcl-2 and expose its proapoptotic BH3 domain. *J. Exp. Med.* 205, 1029-36.

Tosa, N., A. Iwai, T. Tanaka, T. Kumagai, T. Nitta, S. Chiba, M. Maeda, Y. Takahama, T. Uede and T. Miyazaki (2010) Critical function of death-associated protein 3 in T cell receptor-mediated apoptosis induction. *Biochem. Biophys. Res. Commun.* 395, 356-60.

Tosa, N., M. Murakami, W. Y. Jia, M. Yokoyama, T. Masunaga, C. Iwabuchi, M. Inobe, K. Iwabuchi, T. Miyazaki, K. Onoe, M. Iwata and T. Uede (2003) Critical function of T cell death-associated gene 8 in glucocorticoid-induced thymocyte apoptosis. *Int. Immunol.* 15, 741-9.

Tronche, F., C. Opherk, R. Moriggl, C. Kellendonk, A. Reimann, L. Schwake, H. M. Reichardt, K. Stangl, D. Gau, A. Hoeflich, H. Beug, W. Schmid and G.

Schutz (2004) Glucocorticoid receptor function in hepatocytes is essential to promote postnatal body growth. *Genes Dev.* 18, 492-7.

Tuckermann, J. P., H. M. Reichardt, R. Arribas, K. H. Richter, G. Schutz and P. Angel (1999) The DNA binding-independent function of the glucocorticoid receptor mediates repression of AP-1-dependent genes in skin. *J. Cell Biol.* 147, 1365-70.

Vacca, A., M. P. Felli, A. R. Farina, S. Martinotti, M. Maroder, I. Screpanti, D. Meco, E. Petrangeli, L. Frati and A. Gulino (1992) Glucocorticoid receptor-mediated suppression of the interleukin 2 gene expression through impairment of the cooperativity between nuclear factor of activated T cells and AP-1 enhancer elements. *J. Exp. Med.* 175, 637-46.

Vacchio, M. S. and J. D. Ashwell (2000) Glucocorticoids and thymocyte development. *Semin. Immunol.* 12, 475-85.

Villunger, A., E. M. Michalak, L. Coultas, F. Mullauer, G. Bock, M. J. Ausserlechner, J. M. Adams and A. Strasser (2003) p53- and drug-induced apoptotic responses mediated by BH3-only proteins puma and noxa. *Science.* 302, 1036-8.

Wang, A., J. Rud, C. M. Olson, Jr., J. Anguita and B. A. Osborne (2009) Phosphorylation of Nur77 by the MEK-ERK-RSK cascade induces mitochondrial translocation and apoptosis in T cells. *J. Immunol.* 183, 3268-77.

Wang, Z., M. H. Malone, H. He, K. S. McColl and C. W. Distelhorst (2003a) Microarray analysis uncovers the induction of the proapoptotic BH3-only protein Bim in multiple models of glucocorticoid-induced apoptosis. *J. Biol. Chem.* 278, 23861-7.

Wang, Z., M. H. Malone, M. J. Thomenius, F. Zhong, F. Xu and C. W. Distelhorst (2003b) Dexamethasone-induced gene 2 (dig2) is a novel pro-survival stress gene induced rapidly by diverse apoptotic signals. *J. Biol. Chem.* 278, 27053-8.

Wei, M. C., W. X. Zong, E. H. Cheng, T. Lindsten, V. Panoutsakopoulou, A. J. Ross, K. A. Roth, G. R. MacGregor, C. B. Thompson and S. J. Korsmeyer (2001) Proapoptotic BAX and BAK: a requisite gateway to mitochondrial dysfunction and death. *Science.* 292, 727-30.

Woodward, M. J., J. de Boer, S. Heidorn, M. Hubank, D. Kioussis, O. Williams and H. J. Brady (2010) Tnfaip8 is an essential gene for the regulation of glucocorticoid-mediated apoptosis of thymocytes. *Cell Death Differ.* 17, 316-23.

Woronicz, J. D., B. Calnan, V. Ngo and A. Winoto (1994) Requirement for the orphan steroid receptor Nur77 in apoptosis of T-cell hybridomas. *Nature*. 367, 277-81.

Woronicz, J. D., A. Lina, B. J. Calnan, S. Szychowski, L. Cheng and A. Winoto (1995) Regulation of the Nur77 orphan steroid receptor in activation-induced apoptosis. *Mol. Cell Biol*. 15, 6364-76.

Wyllie, A. H. (1980) Glucocorticoid-induced thymocyte apoptosis is associated with endogenous endonuclease activation. *Nature*. 284, 555-6.

Zhang, Y., S. Rosenberg, H. Wang, H. Z. Imtiyaz, Y. J. Hou and J. Zhang (2005) Conditional Fas-associated death domain protein (FADD): GFP knockout mice reveal FADD is dispensable in thymic development but essential in peripheral T cell homeostasis. *J. Immunol*. 175, 3033-44.

In: Nuclear Receptors
Editors: M. K. Bates and R. M. Kerr
ISBN: 978-1-61209-980-4

Chapter 4

NITRIC OXIDE: A TOOL TO BLOCK NUCLEAR RECEPTOR ACTIVITY

Klaus-Dieter Spindler, Martin Laschak and Marcus V. Cronauer*
Institut für Allgemeine Zoologie und Endokrinologie,Universität Ulm, D-89081 Ulm, Albert-Einstein-Allee 11

INTRODUCTION

Nitric oxide (NO), a free radical gas, is an omnipresent intercellular messenger in all vertebrates. Originally described as a cardiovascular signal molecule (Ignarro, 1989) NO elicites a variety of physiological functionslike muscle contractility, platelet aggregation, metabolism, neuronal activity, and immune responses in a broad range of tissues.

The molecule originates from the action of nitric oxide synthases (NOS) which are either induced (iNOS) or constitutively expressed (eNOS, nNOS). The underlying mechanisms of NO action are primarily an elevation of guanosine 3',5'-cyclic monophosphate due to the stimulation of soluble guanylyl cyclase, inhibition of mitochondria respiration and nitrosylation of proteins (Pacher et al., 2007; Gao, 2010).

* Universitätsklinikum Ulm, Klinik für Urologie und Kinderurologie, Urologisches Forschungslabor, Prittwitzstr. 43, D-89075 Ulm

In vitro, NO has been shown to modulate the activity of a variety of nuclear receptors (NR) of the steroid receptor superfamily like the estrogen (ER) or the androgen receptor (AR) (Garban 2005, Cronauer 2007).

Nuclear receptors are transcription factors characterized by a ligand binding domain, a highly conserved DNA binding domain consisting of two Cys4-type zinc fingers and a transactivation domain, possessing one of the two activation function domains.

Upon ligand binding, NR predominantly act via binding to so called hormone response elements on the DNA thus regulating gene expression. Essential features which regulate the activity and function of NR are receptor concentration, post translational modifications like phosphorylation or acetylation which trigger receptor dimerization, nucleocytoplasmic shuttling or binding of comodulators.

Despite their different modes of action, the signaling pathways of nitric oxide and NR interfere a manifold. Estrogen and progesterone are known to up-regulate NO synthesis whereas glucocorticoids and progesterone decrease NO bioavailability (reviewed in: Duckles and Miller, 2010). Due to its unique physicochemical properties (high reactivity of NO-radicals, short half-life, excellent membrane permeability) the NO-molecule is also able to directly interact with nuclear receptors thereby blocking their activity.

In the present review we will summarize the so far known effects of NO on nuclear receptors and demonstrate its potential use as a tool for studying gene regulation. In addition we will discuss the physiological relevance of NO/NR interaction.

Nitric-Oxide Delivering Compounds

In order to reach supraphysiological levels of NO needed in experimental or therapeutical studies, a variety of NO-releasing compounds has been synthesized (Burgaud et al., 2002, Figure 1). The majority of NO-releasing compounds are either belonging to the group of diazeniumdiolates, which allow the most predictable release of nitric oxide or to the group of S-nitrosothiols, where NO release is dependent on a variety of parameters like free SH-groups, Cu^{1+} and Fe^{2+} concentrations.

Irrespective of the chemical class, a series of photosensitive precursors to nitric oxide have been synthesized. These compounds allow a controlled photochemical release of NO (Ruane et al., 2002; Pavlos et al., 2004) and are useful tools in photodynamic therapy (Pavlos et al., 2005).

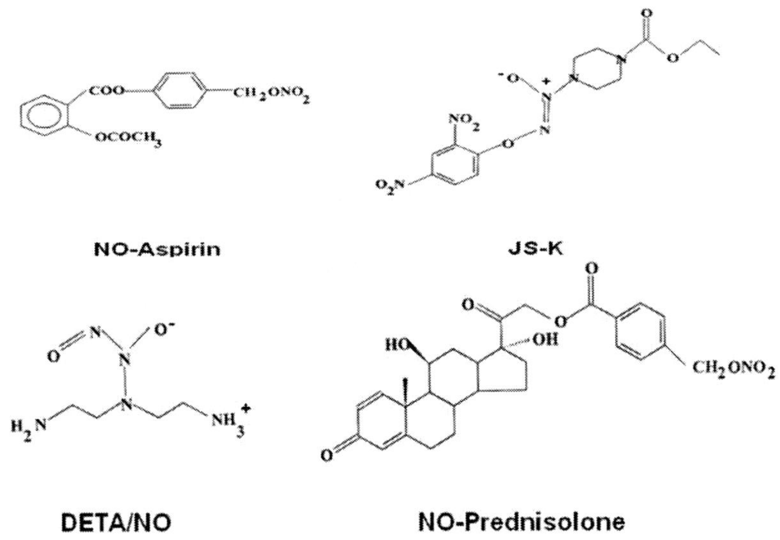

NO-Aspirin JS-K

DETA/NO NO-Prednisolone

Figure 1. Chemical structure of four representatives of NO-releasing compounds.

The half-life as well as the release of NO of different NO-releasing compounds exhibit a large variability even within members of the same chemical class. For example, the commonly used diazeniumdiolate DETA/NO ((Z)-1-[N-(2-Aminoethyl)-N-(2-ammonioethyl)amino]diazen-1-ium-1,2-diolat) releases nitric oxide spontaneously in aqueous solutions at 37°C with a half-life of DETA/NO of eight hours (Hrabie et al., 2000; Berendji et al., 1997), whereas JS-K (*O*^{2}-(2,4-Dinitrophenyl)-1-[(4-ethoxycarbonyl)piperazin-1-yl]diazen-1-ium-1,2-diolat), another diazeniumdiolate, releases nitric oxide only after enzymatic action of glutathione S-transferases which form an intermediate complex, which disintegrates with a half-life of eight minutes (Shami et al., 2003).

INTERACTION OF NO WITH NUCLEAR RECEPTORS

a. DNA and hormone binding, transactivation. The physicochemical properties of NO have been summarized by Fukuta and coworkers (Fukuta et al., 2000). Although NO is a radical its reactivity in biological systems is relatively low. Due to its small size and its lipophilicity the NO-molecule easily passes cell membranes. However, NO will most likely react with oxygen, thus explaining its oxidation under physiological conditions. Usual reaction products of NO with oxygen are the so called reactive nitrogen oxide intermediates (RNOI) or higher nitrogen oxides (NO_x). RNOI are highly reactive short lived species exhibiting a much broader spectrum of activity towards biological systems and are for example responsible for the nitrosative stress (Hausladen et al., 1996).There are many NO-related chemical reactions leading to stable or unstable products. In general the type of reaction strongly depends on the local concentration of NO which primarily determines whether NO acts as a signal molecule or induces nitrosative stress (Kröncke, 2001). Molecular targets for this stress reaction are free thiol groups which are nitrosylated to R-SNO. One of the characteristics of NR is the presence of two zinc fingers in the DNA binding domain, which are mandatory for DNA binding and the final action of steroid or thyroid hormones. These zinc fingers are formed by 4 cysteines complexed by Zn^{2+} (Figure 2). As a model substance to study the interaction of NO with zinc-sulfur cluster structures, methallothionein has been investigated. It was clearly shown that NO releases Zn^{2+} from this substance concomitant with a decrease of free SH-groups (Kröncke et al., 1994; Kröncke, 2001). Subsequently it has been shown that in nuclear receptors indeed NO leads to a disruption of receptor DNA interaction and finally to a loss of function of these receptors (Table 1). This is demonstrated exemplarily for the androgen receptor where NO disrupts DNA binding of the AR thereby diminishing transactivation (Figure 3). Interestingly the inhibition of the NR-DNA interaction is reversible (Kröncke and Carlberg, 2000). Since transcription factors can also act as repressors, it is evident that in these cases destruction of the zinc finger structure and loss of binding to DNA leads to an activation of transcription, as shown by a decreased binding of transcription factor SP-1 to the promoter of tumor necrosis factor α (TNFα) (Wang et al., 1999). Moreover zinc fingers built by two cysteines and two histidines as in the early growth response factor 1 (EGR-1) are affected by nitric oxide in the same way as cys_4 zinc fingers of nuclear receptors (Kröncke, 2001). In addition to the SH-groups of zinc finger cysteines, there are additional free SH groups in NR that are important for receptor function.This has been demonstrated for the phylogenetically old ecdysteroid receptor where these SH-groups are important for hormone binding and uptake (Londershausen and Spindler, 1981; Daig and Spindler, 1983; Dinan and Spindler, 1986). It is therefore not surprising that NO is also involved in hormone binding, as demonstrated for the glucocorticoid receptor (Table 1).

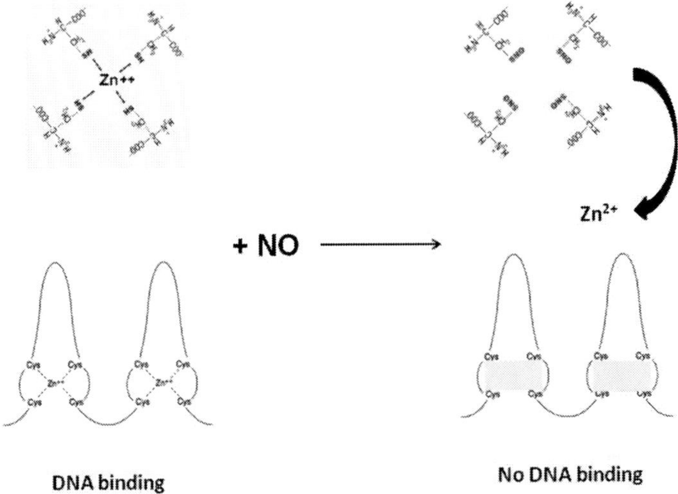

Figure 2. Effect of NO on zinc fingers from nuclear receptors.

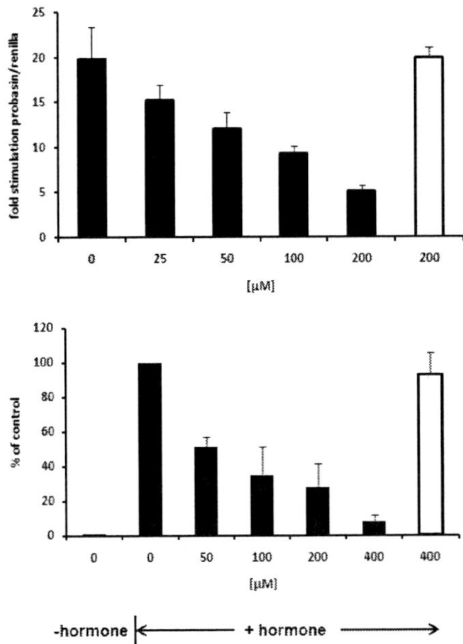

Figure 3. NO dose dependently inhibits the DNA-binding activity of the AR (a) and reporter gene activity (b). Black bars: DETA/NO, white bars: DETA; means ± standard deviations, n = 4. (Redrawn from Cronauer et al., Oncogene 26 (2007) 1875-1884; with the kind permission of the publisher).

Table 1. Effects of nitric oxid on nuclear receptors

Receptors	Effects	Literature
Ecdysteroid receptor/ ultraspiracle (EcR/Usp)	When studied separately, nuclear localization of neither Usp nor EcR is reduced; the hormone induced increase in nuclear localization of EcR and of the heterodimer is prevented by NO	Cronauer et al., 2007b
Androgen receptor (AR)	Dose-dependent reduction of DNA binding of the androgen receptor, reduction of transactivation in prostate cancer cell lines. Inhibition of androgen dependent cell lines by NO more efficient than in androgen receptor-negative cell lines	Cronauer et al., 2007a
Estrogen receptor α	Reversible reduction of DNA binding, transactivation reduced, S-nitrosylation of the ER; selective inhibition of estrogen-induced gene expression without affecting nongenomic events	Garban et al., 2005; Marino et al., 2001
Vitamin D receptor/retinoic acid receptor (VDR/RXR)	Reversible reduction of DNA binding to hormone response elements. Reduction of transactivation by DETA-NO (Ic_{50}-value: 0.5 mm), inhibition of dimerisation	Kröncke and Carlberg, 2000
Glucocorticoid receptor (GR)	Inhibition of hormone binding by NO, prevented by simultaneous application of DDT	Galigniani et al., 1999

b. Receptor stability and nuclear transport. Since nitrosative stress might reduce protein synthesis and/or stimulate protein degradation the question arises whether NO is able to influence intranuclear receptor levels. To our knowledge this has only been studied for the ecdysteroid receptor from *Drosophila melanogaster* (Cronauer et al., 2007b) and the human androgen receptor (Cronauer et al., 2007a). Whereas treatment of prostate cancer cells with DETA/NO did not alter intracellular AR-levels, the concentration of both heterodimerization partners of the ecdysteroid receptor EcR and Usp were diminished when treated under the same conditions. Since different cell lines were used in the above mentioned studies it cannot be ruled out that the differing stability of the AR and the ecdysone receptor is an inherent feature of the corresponding receptors or due to a different sensitivity of the cell lines towards NO. The import of nuclear receptors is a prerequisite for the regulation of hormone induced gene expression. In the absence of hormonal stimuli NO is unable to influence the import of EcR,Usp and the AR. Interestingly NO does not inhibit the hormone induced import of the AR, but reduces partially the ligand induced import of the ecdysteroid receptor. Again, the reasons for this different behaviour are unknown. Future work will be focussing on a comparative analysis of stability and nuclear transport of different nuclear receptors.

c. Dimerisation. Dimerization of nuclear receptors, either as homodimers or as heterodimers between different partners, is mandatory for steroid/ thyroid hormone or Vitamin A/D action. For example, the vitamin D receptor (VDR) can dimerize with the retinoic acid receptor (RXR). In this case it has been shown by *in vitro* experiments that dimerization is inhibited by NO. Each of the two receptors is sensitive towards NO. However, DNA bound heterodimers as well as ligand bound heterodimers exhibited a lower sensitivity towards NO induced stress (Kröncke and Carlberg, 2000). In co-immunoprecipitation experiments an inhibition of protein-protein interactions by NO has been shown for β-catenin/T-cell factor(TCF) in colon cancer cells (Nath et al., 20003) and in a bacterial system, for cochaperone/chaperon interaction (Kröncke et al, 2001).

d. Effect of NO on proliferation and cytotoxicity. In contrast to NO, which is of relatively low reactivity despite being a radical, the reactive nitrogen oxide intermediates are highly active in biological systems, as demonstrated by the ability of peroxynitrite to damage DNA(Niles et al., 2006). Therfore it is thus not surprising that locally increased levels of NO produced by activated macrophages do not only kill bacteria and other parasites (Bogdan et al., 2000; Nathan and Shiloh, 2000) but also pancreatic islet cells (Kröncke et al., 1991). Because of serious differences of various cell types in their sensitivity towards NO and its oxidation products (Kröncke et al., 1993) one has to test the effects of NO on proliferation and cytotoxicity for each cell type. In human prostate cancer cell lines it has been demonstrated that NO-delivering substances inhibit proliferation and induce apoptosis (Huguenin et al, 2004; Royle et al., 2004). However, the effect is much more pronounced in AR-positive cells as compared to AR-negative ones (Huguenin et al., 2004; Cronauer et al., 2007a). Although the NO-donating substance JS-K is highly toxic towards human multiple myeloma cells *in vitro*, the same substance is relatively well tolerated when used ~~a~~ in myeloma xenograft mouse model (Kiziltepe et al., 2007).

e. Nitric oxide, zinc release and protein S-nitrosylation. Nitric oxide disrupts the structure of the zinc fingers by nitrosating free SH-groups the latter leading to a release of Zn^{2+} from zinc finger structures. Increased cytoplasmic or nuclear free Zn^{2+} can interfere with various signaling pathways, involving for example phosphoinositide 3'-kinases and the Ser/Thr protein kinase Akt, resulting in an inactivation of transcriptional regulators of the FoxO family. Zn^{2+} dyshomeostasis and the resulting human disorders or diseases were summarized recently (Kröncke and Klotz, 2009). A further aspect of NO action is its potency to nitrosate not only free SH-groups but also tyrosine residues in proteins. A review highlighting the importance of protein S-nitrosylation for various pathophysiological processes has been recently published (Foster et al., 2009). The fact that nitrotyrosine content positively correlates with the expression of inducible nitric oxide synthase and thus with endogenous nitric oxide has been recently shown for prostate cancer tissue (Cronauer et al., 2007a). In addition the same authors also demonstrated that prostate cancer cells incubated with nitric oxide donors show a positive nitrotyrosine-staining.

Indirect Effects of Nitric Oxide on Nuclear Receptor Signaling

Nitric oxide has a variety of downstream effects, involving WNT, NF-KB, MAPK, NOS and *Nrf2* pathways. An example for the influence of NO on the canonical WNT pathway is presented here. A crucial step in the canonical WNT-pathway is the control of intracellular β-catenin levels. Under normal conditions increasing β-catenin levels lead to the activation of T-Cell factor/lymphoid enhancer factor (TCF/LEF). The β-catenin/TCF complex subsequently binds to the DNA, thereby inducing the transcription of target genes in different cell types like for example in colon cancer cells (Nath et al., 2003) or in the AR-positive 22Rv1 prostate cancer cells (Cronauer et al., 2005). Moreover, a pronounced cross talk between WNT und AR signaling pathways including a direct binding of β-catenin to the activation function 2 (AF2) of AR has been described (Beildeck and Gelman, 2010; Song and Gelman, 2005, Cronauer et al. 2005).

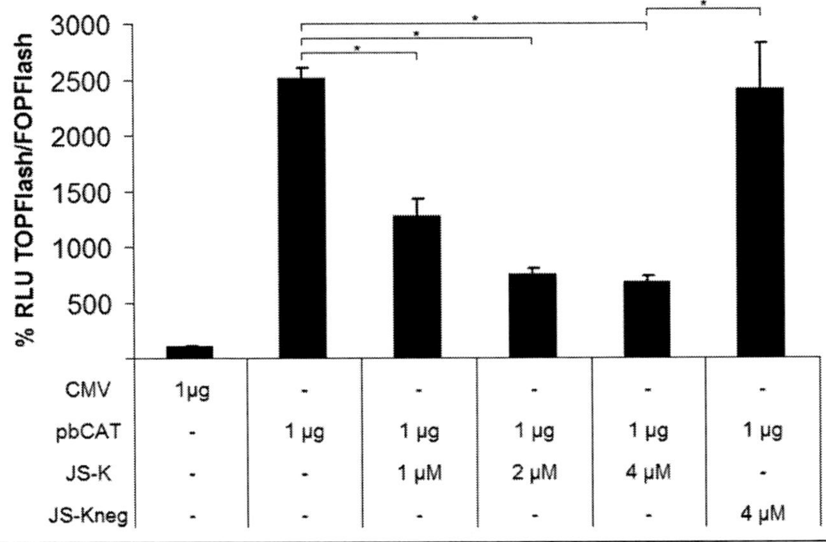

Figure 4. Influence of JS-K on the transactivation of TCF. 22Rv1-cells were cotransfected with pTOPFlash (TOP) and pRL-TK (Ren) or with pFOPFlash (FOP) and pRL-TK. As a control the empty vector pCMV-NeoBam (CMV) was cotransfected or the vector coding for S33Y β-catenin (pbCAT). Cells were incubated with JS-K or an inactivated form of JS-K termed JS-Kneg for 44 hours. M + SD, n = 3; * $P<0.01$.

In 22Rv1 cells the NO-Donor JS-K was recently shown to dose dependently inhibit TCF-induced transactivation (Figure 4; own, unpublished results). Since WNT signaling is an important feature in prostate cancer (Verras et al., 2004; Yardy and Brewster, 2005; Terry et al., 2006) the modulation of WNT-signaling by NO might be a new target for a therapeutical intervention (Rinnab et al., 2008).

Applications of Nitric Oxide for Therapeutic Use

Nitric-oxide releasing compounds are widely used forthe treatment and prevention of cancer and various other malignancies (Rigas, 2007; Shami et al., 2006; Findly et al., 2004; Mocellin et al., 2007; Hirst and Robson, 2010a, b; Hickok and Thomas, 2010; Ma et al., 2010; Burgaud et al., 2002; Pavlos et al., 2005; Kanwar et al. 2009; Ahmad et al., 2009). Promising results have been described for breast cancer (Pervin et al., 2010), pancreatic tumors (Wang and Xie, 2010) and tumors of the nervous tissue (Peña-Altamira et al., 2010) whereas conflicting results have been reported in melanomas (Ekmekcioglu et al.,

2005).These differences might be due to the fact that nitric oxide displays a biphasic behavior in cancer biology (Ridnour et al., 2006).

Future Perspectives

Because of the manifold targets for NO two serious problems for all applications arise: first, the high degree of toxicity of NO and its oxidative reaction products and second, connected to the first problem, the targeting of the drugs. One attempt to solve both problems at least in case of nuclear receptors is to couple an NO-donating moicty to the corresponding steroid hormone thus allowing a very specific targeting and an intracellular release of NO at lower concentrations. One successful approach has been made with the synthesis of NCX-1015, a nitric oxide releasing derivative of prednisolone (Paul-Clark et al., 2000). This compound has been proven to be effective in the treatment of an experimental arthritis (Paul-Clark et al., 2002) and against an experimentally induced colitis in mice (Fiorucci et al., 2002). One future development could be the synthesis of an NO releasing androgen/anti-androgen for the treatment of advanced prostate cancer.

References

Ahmad, R; Rasheed, Z, Ahsan, H (2009) Biochemical and cellular toxicology of peroxynitrite: implications in cell death and autoimmune phenomenon. *Immunopharmacol Immunotoxicol 31*, 388-396.

Beildeck, ME; Gelmann, EP, Byers, SW (2010) Cross-regulation of signaling pathways: an example of nuclear hormone receptors and the canonical Wnt pathway *Exp. Cell Res. 316*, 1763-72.

Berendji, D; Kolb-Bachofen, V; Meyer, KL; Grapenthin, O; Weber, H; Wahn, V, Kröncke, KD (1997). Nitric oxide mediates intracytoplasmic and intranuclear zinc release. *FEBS Lett. 405*, 37-41.

Bogdan, C; Röllinghoff, M, Diefenbach, A (2000) Reactive oxygen and reactive nitrogen intermediates in innate and specific immunity. *Curr. Opin. Immunol. 12*, 64-76.

Burgaud, JL; Riffaud, JP, Del Soldato, P (2002) Nitric-oxide releasing molecules: A new class of drugs with several major indications. *Curr. Pharm. Design. 8*, 201-213.

Cronauer, MV; Schulz, WA; Ackermann, R, Burchardt, M (2005). Effects of WNT/beta-catenin pathway activation on signaling through T-cell factor and androgen receptor in prostate cancer cell lines. *Int. J. Oncol. 26*, 1033-1040.

Cronauer, MV; Ince, Y; Engers, R; Rinnab, L; Weidemann, W; Suschek, CV; Burchardt, M; Kleinert, H; Wiedenmann, J; Sies, H; Ackermann, R, Kröncke, KD (2007a). Nitric oxide-mediated inhibition of androgen receptor activity: possible implications for prostate cancer progression. *Oncogene 26*, 1875-1884.

Cronauer, MV; Braun, S; Tremmel, Ch; Kröncke, KD, Spindler-Barth M (2007b). Nuclear localization and DNA binding of ecdysone receptor and ultraspiracle. *Arch. Insect. Biochem. Physiol. 65*, 125-133.

Daig, K, Spindler, K-D (1983) Uptake and retention of moulting hormones by the integument of crayfishes in vitro. II. Influence of metabolic inhibitors and sulfhydryl-group inhibitors. *Mol. Cell Endocrinol. 32*, 73-79.

Dinan, L, Spindler, K-D (1986) The use of mercurial agents in the study of mode of action of ecdysteroids. *Insect Biochem. 16*, 135-141.

Duckles, SP, Miller, VM (2010) Hormonal modulation of endothelial NO production. *Pflügers Arch. 459*, 841-851.

Ekmekcioglu, S; Tang, CH, Grimm, EA (2995). NO news is not necessarily good news in cancer, *Curr. Cancer Drug Targets, 5*, 103-115.

Findlay, VJ; Townsend, DM; Saavedra, JE; Buzard, GS; Citro, ML; Keefer, LE; Ji, X, Tew, KD (2004). Tumor cell response to a novel glutathione S-transferase-activated nitric oxide-releasing prodrug. *Mol. Pharmacol. 65*, 1070-1079.

Fiorucci, S; Antonelli, E; Distrutti, E; DelSoldato, P; Flower, RJ; Clark, MJP; Morelli, A; Perretti, M, Ignarro, LJ (2002) NCX-1015, a nitric-oxide derivative of prednisolone, enhances regulatory T cells in the lamina propria and protects against 2,4,6-trinitrobenzene sulfonic acid-induced clitis in mice. *Proc. Natl. Acad. Sci. USA 99*, 15770-15775.

Foster, MW; Hess, DT, Stamler, JS (2009) Protein S-nitrosylation in health and disease: a current perspective. *Trends Mol. Med. 15*, 391-404.

Fukuta, JM; Cho, JY, Switzer, CH (2000) The chemical properties of nitric oxide and related nitrogen oxides. In: Ignarro, LJ. *Nitric Oxide: Biology and Pathobiology,* Academic Press, San Diego.

Galigniana, MD; Piwien-Pilipuk G, Assreuy I (1999) Inhibition of glucocorticoid receptor binding by nitric oxide. *Mol. Pharmacol. 55*, 317-323.

Gao Y (2010) The multiple actions of NO. *Pflügers Arch. 459*, 829-839.

Garban, HJ; Márquez-Garba, DC: Pietras, RJ, Ignarro, LJ (2005) Rapid nitric oxide-mediated S-nitzrosylation of estrogen receptor: Regulation of estrogen-dependent gene transcription. *Proc. Natl. Acad. Sci. USA 102*, 2632-2636.

Hausladen, A; Privalle, CT; Keng, T; DeAngelo, J, Stamler, JS (1996) Nitrosative stress: activation of the transcription factor OxyR. *Cell 86*, 719-729.

Hickok, JR, Thomas, DD (2010) Nitric oxide and cancer therapy: the emperor has NO clothes. *Curr. Pharm. Des. 16*, 381-391.

Hirst, D, Robson, T (2010a) Nitric oxide in cancer therapeutics: interaction with cytotoxic chemotherapy. *Curr. Pharm. Des. 16*, 411-420.

Hirst, DG, Robson, T (2010b) Nitrosative stress as a mediator of apoptosis: implications for cancer therapy. *Curr. Pharm. Des. 16*, 45-55.

Hrabie, JA; Arnold, EV; Citro, ML; George, C, Keefer, LK (2000) Reaction of nitric oxide at the beta-carbon of enamines. A new method of preparing compounds containing the diazeniumdiolate functional group. *J. Org. Chem. 65*, 5745-5751.

Huguenin, S; Fleury-Feith, J; Kheuang, L; Jaurand,M-C; Bolla, M, Riffaud, J-P (2004) Nitrosulindac (NCX1102):A new nitric oxide-donating non-steroidal anti-inflammatory drug (NO-NSAID), inhibits proliferation and induces apoptosis in human prostatic epithelial cell lines.*The Prostate 61*, 132-141.

Ignarro, LJ (1989). Endothelium-derived nitric oxide: actions and properties. *FASEB J. 3,* 31-36.

Kanwar, JR; Kanwar, RK; Burrow, H, Baratchi, S (2009) Recent advances on the roles of NO in cancer and chronic inflammatory disorders. *Curr. Med. Chem. 16*, 2373-2394.

Kröncke, KD (2001) Zinc finger proteins as molecular targets for nitric oxide-mediated gene regulation. *Antioxid. Redox. Signal. 3*, 565-575.

Kröncke, KD, Carlberg, C (2000). Inactivation of zinc finger transcription factors provides a mechanism for a gene regulatory role of nitric oxide. *FASEB J. 14*, 166-173.

Kröncke, KD; Fehsel, K, Schmidt; T, Zenke; FT, Dasting; I, Wesener; JR, Bettermann; H, Breunig, KD, Kolb-Bachofen V (1994). Nitric oxide destroys zinc-sulfur clusters inducing zinc release from metallothionein and inhibition of the zinc finger-type yeast transcription activator LAC9.*Biophys. Biochem. Res. Comm. 200,* 1105-1110.

Kröncke, KD; Haase, H; Beyersmann, D; Kolb-Bachofen, V, Hayer-Hartl, MK (2001) NNitric oxide inhibits the cochaperone activity of the RING finger-like protein DnaJ: *Nitric. Oxide, 5*, 289-295.

Kröncke, KD; Kolb-Bachofen, V, Berschik, B; Burkart, V, Kolb, H (1991) Activated macrophages kill pancreatic syngeneic islet cells via arginine-dependent nitric oxide generation. *Biochem. Biophys. Res. Commun. 175*, 752-758.

Kröncke, KD; Brenner, HH; Rodriguez, ML; Etzkom, K; Noack, EA; Kolb, H, Kolb-Bachofen, V (1993) Pancreatic islet cells are highly susceptible towards the cytotoxic effects of chemically generated nitric oxid. *Biochim. Biophys. Acta, 1182*, 221-229.

Kröncke, KD, Klotz, LO (2009) Zinc fingers as biologic redox switches? *Antioxid. Redox. Signal, 11*, 1015-1027.

Londershausen, M; Spindler, K-D (1981) Characterization of cytoplasmic ecdysteroid receptors in the hypodermis of the crayfish, *Orconectes limosus. Mol. Cell Endocrinol. 24*, 253-265.

Ma, Q; Wang ,Z; Zhang, M; Hu, H; Li, J; Zhang, D; Guo, K, Sha, H (2010) Targeting the L-arginine-nitric oxide pathway for cancer treatment. *Curr. Pharm. Des. 16*, 392-410.

Marino, M; Ficca, R; Ascenzi, P, Trentalance, A (2001) Nitric oxide inhibits selectively the 17beta-estradiol-induced gene expression without affecting nongenomic events in HeLa cells, *Biochem. Biophys. Res. Commun. 286*, 529-533.

Mocellin, S; Bronte, V, Nitti, D (2007). Nitric oxide, a double edged sword in cancer biology: searching for therapeutic opportunities. *Med. Res. Rev. 27*, 317-352.

Nath, N; Kashfi, K; Chen, J, Rigas, B (2003) Nitric oxide-donating aspirin inhibits ß-catenin/T cell factor (TCF) signaling in SW480 colon cancer cels by disrupting the nuclear ß-catenin-TCF association. *Proc. Natl. Acad. Sci. USA 100*, 12584-12589.

Nathan, C., Shiloh, MU (2000) Reactive oxygen and nitrogen intermediates in the relationship between mammalian hosts and microbial pathogens. *Proc. Natl. Acad. Sci. USA 97*, 8841-8848.

Niles, JC; Wishnok, JS, Tannenbaum, SR (2006) Peroxynitrite-induced oxidation and nitration products of guanine and 8-oxoguanine: structures and mechanisms of product formation. *Nitric. Oxide, 14*, 109-121.

Pacher, P; Beckman, JS; Liaudet, L (2007). Nitric oxide and peroxynitrite in health and disease. *Physiol. Rev. 87*, 315-424.

Paul-Clark, M; Del Soldato, P; Fiorucci, S; Flower, RJ, Perretti, M (2000) 21-NO-prednisolone is a novel nitric oxide-releasing derivative of prednisolone with enhanced anti-inflammatory properties. *BR. J. Pharmacol. 131*, 1345-1354.

Paul-Clark, M; Mancini, L; Del Soldato, P; Flower, RJ, Perretti, M (2002) Potent antiarthritic properties of a glucocorticoid derivative, NCX-1015, in an experimental model of arthritis. *Proc. Natl. Acad. Sci. USA, 99*, 1677-1682.

Pavlos, CM; Xu, H, Toscano, JP (2004) Controlled photochemical release of nitric oxide from O^2-substituted diazeniumdiolates. *Free Rad. Biol. Med. 37*, 745-754.

Pavlos, CM; Xu, H, Toscano, JP (2005) Photosensitive precursors to nitric oxide. *Curr. Top. Med. Chem. 5*, 637-647.

Peña-Altamira, E; Petazzi, P, Contestabile, A (2010) Nitric oxide control of proliferation in nerve cells and in tumor cells of nervous origin. *Curr. Pharm. Des. 16*, 440-450.

Pervin, S; Chaudhuri, G, Singh, R (2010) NO to breast: when, why and why not? *Curr. Pharm. Des. 16,* 451-462.

Rigas, B (2007). Novel agents for cancer prevention based on nitric oxide. *Biochem. Soc. Trans. 35*, 1364-1368.

Ridnour, LA; Thomas, DD; Donzelli, S; Espey, MG; Roberts, DD; Wink, DA, Isenberg, JS (2006) The biphasic nature of nitric oxide responses in tumor biology. *Antioxid. Redox. Signal,* 8, 1329-1337.

Rinnab, L; Hessenauer, A; Schütz, SV; Schmid, E; Küfer, R; Finter, F; Hautmann, RE; Spindler, KD, Cronauer MV (2008). Role of androgen receptors in hormone-refractory prostate cancer: molecular basics and experimental therapy approaches. *Urologe A 47*, 314-325.

Royle JR; Ross, JA; Ansell, I; Bollina, P; Tulloch, DN, Habib, FK (2004) Nitric oxid donating nonsteroidal anti-inflammatory drugs induce apoptosis in human prostate cancer cell systems and human prostatic stroma via caspase-3. *J. Urol. 172*, 338-344.

Ruane, PH; Bushan, KM; Pavlos, CM; Raechelle, A; D'Sa, Toscana, JP (2002) Controlled photochemical release of nitric oxide from O^2-benzoyl-substituted diazeniumdolates. *J. Am. Chem. Soc. 124*, 9806-9811.

Shami, PJ; Saavedra, JE; Boninfant, CL; Ji, X; Keefer, LK; Chu, J; Udupi, V; Malaviya, S; Carr, BJ; Kar, S; Wang, M; Jia, L; Ji, X, Keefer, LK (2006) Antitumor activity of JS-K [O2-(2,4-Dinitrophenyl)1-[(4-Etholxycarbonyl)piperazin-1-ium-1,2diolate] and related O2-Aryl-diazeniumdiolates in vitro and in vivo, *J. Med. Chem. 49***,** 4356-4366.

Shami, PJ; Saavedra, JE; Wang, LY; Bonifant, CL; Diwan, BA; Singh, SV; Gu, Y; Fox, SD; Buzard, GS; Citro, ML; Waterhouse, DJ; Davies, KM; Ji, X, Keefer, LK (2003) JS-K, a glutathione/glutathione S-transferase-activated nitric oxide donor of the diazeniumdiolate class with potent antineoplastic activity. *Mol. Cancer Ther. 2*, 409-417.

Song, LN, Gelmann, EP (2005). Interaction of beta-catenin and TIF2/GRIP1 in transcriptional activation by the androgen receptor. *J. Biol. Chem. 280*, 37853-37867.

Wang, L, Xie, K (2010), Nitric oxide and pancreatic cancer pathogenesis, prevention, and treatment. *Curr. Pharm. Des. 16*, 421-427.

Terry, S; Yang, X; Chen, MW; Vacherot, F, Buttyan, R (2006). Multifaceted interaction between the androgen and Wnt signaling pathways and the implication for prostate cancer. *J. Cell Biochem. 99*, 402-410.

Wang, S; Wang, W; Wesley, RA, Danner RL (1999) A Sp1 binding site of the tumor necrosis factor a promoter functions as a nitric oxide response element. *J. Biol. Chem. 274*, 33190-33193.

Yardy, GW, Brewster, SF (2005) Wnt signalling and prostate cancer. *Prostate Cancer Prostatic Dis. 8*, 119-126.

In: Nuclear Receptors
Editors: M. K. Bates and R. M. Kerr
ISBN: 978-1-61209-980-4

Chapter 5

VITAMIN K AS A LIGAND OF STEROID AND XENOBIOTIC RECEPTOR

Kotaro Azuma[1], Kuniko Horie-Inoue[2], Yasuyoshi Ouchi[1] and Satoshi Inoue[1,2]
[1]University of Tokyo
[2]Saitama Medical University, Japan

Vitamin K is a fat-soluble vitamin essential for blood coagulation. Natural vitamin K includes vitamin K1 (phylloquinone), present mainly in vegetables, and vitamin K2 (menaquinone), which is synthesized by microorganisms and is present in food such as natto (fermented soy beans). Vitamin K1 is converted to vitamin K2, the functionally active form, in the body [1].

Vitamin K was shown to play an essential role in the hepatocytes by maintaining the activity of coagulation factors II, VII, IX, and X and of anticoagulants, protein C, and protein S. Recently, extrahepatic actions of vitamin K have also been reported.

The administration of vitamin K was shown to prevent bone fracture [2, 3], and this led to its clinical application in cases of osteoporosis in East Asian countries. Epidemiological studies have shown that the lack of vitamin K causes osteoarthritis [4] and imposes a risk of coronary artery disease [5]. Some clinical studies have suggested its antitumor activity in cases of hepatocellular carcinoma [6-8] and other cancers [9].

Until recently, vitamin K was primarily known as a co-enzyme of γ-glutamyl carboxylase (GGCX), which converts glutamic acid (Glu) to γ-carboxyglutamic acid (Gla) during the post-transcriptional modification of proteins. During this reaction, vitamin K is oxidized from vitamin K hydroquinone to vitamin K epoxide. Thus far, the following 17 proteins are known to be the substrates of GGCX: coagulation factors II, VII, IX, and X; protein C; protein S; protein Z; nephrocalcin; growth arrest specific-6 (Gas6) [10]; osteocalcin (also known as bone Gla protein [BGP]) [11]; matrix Gla protein (MGP) [12]; periostin [13]; βIg-H3 [13]; proline-rich Gla proteins 1 and 2 [14]; and transmembrane Gla proteins 3 and 4 [15]. Warfarin, which is clinically used as an anticoagulation drug, inhibits the activity of vitamin K epoxide reductase (VKOR). The lack of VKOR activity renders the GGCX-oxidized vitamin K useless, resulting in a decrease in the GGCX activity.

The GGCX activity may be partially responsible for the favorable effect of vitamin K on bone metabolism. One of the single nucleotide polymorphisms of the GGCX gene (8762G>A) is known to cause amino acid substitution (Arg325Gln). GGCX with 325-Gln has higher activity than GGCX with 325-Aln. Elderly people with 325-Gln in both alleles were shown to have higher radial bone mineral density [16].

Moreover, a high concentration of undercarboxylated osteocalcin, which indicates insufficient effect of vitamin K in the bone tissue, was clinically proved to be a risk factor for femoral neck fracture [17]. However, osteocalcin-knockout mice displayed higher bone mineral density than wild-type mice [18]. This implies that the effect of vitamin K in bone tissue is not necessarily conveyed by γ-carboxylation of osteocalcin.

Meanwhile, we unveiled a completely new mechanism vitamin K action. Vitamin K functions as a ligand of the nuclear receptor, Steroid and Xenobiotic Receptor (SXR) (Figure 1). SXR and its murine ortholog Pregnane X Receptor (PXR) (also known as PAR and NR1I2) are nuclear receptors that are activated by various endogenous and dietary substances, pharmaceutical agents, and xenobiotic compounds [19].

Endogenous ligands of SXR/PXR were formerly unknown, and SXR/PXR was classified as an orphan receptor. However, currently, secondary bile acids are known as one of endogenous ligands for these receptors [20, 21]. SXR/PXR is mainly expressed in the liver and intestine [22], where it functions as a xenobiotic sensor by inducing the genes involved in detoxification and drug excretion [22, 23].

Similar to other nuclear receptors, SXR/PXR functions as a ligand-dependent transcription factor. When a ligand binds to SXR/PXR, it forms a heterodimer with RXR; this complex binds to the SXR-responsive element (SXRE) on the promoter or enhancer regions of the target genes.

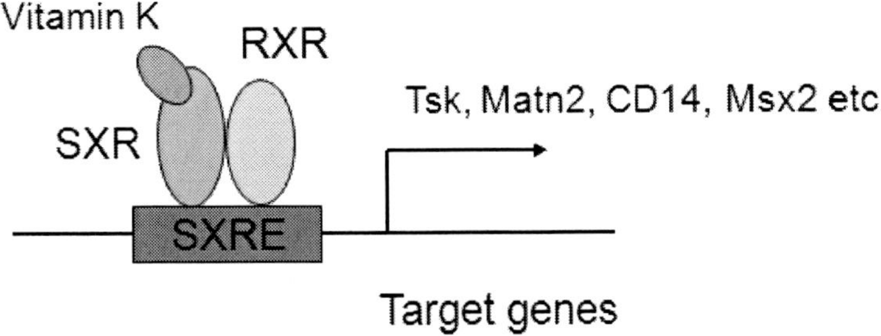

Figure 1. Vitamin K function is mediated by the Steroid and Xenobiotic Receptor (SXR). Vitamin K binds to SXR to form a heterodimer with RXR; this complex then binds to the SXR-responsive element (SXRE) on the promoter or enhancer regions of the target genes. *Tsk* (tsukushi), *Matn2* (Matrilin-2), *CD14*, and *Msx2* are known as SXR target genes that are upregulated by vitamin K treatment in sotioblastic cells.

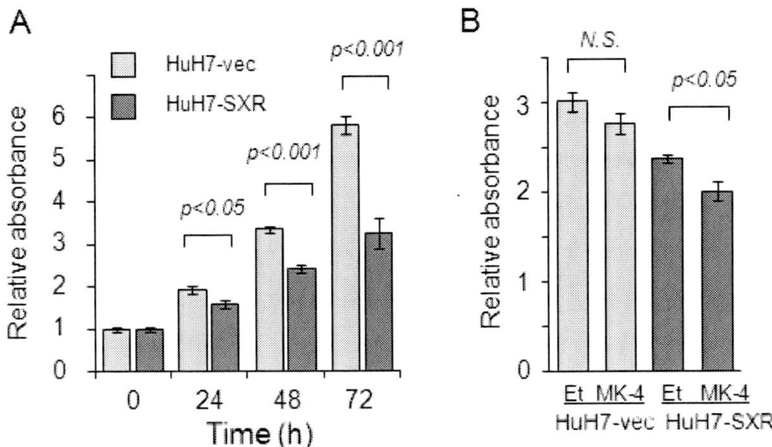

Figure 2. Vitamin K-dependent growth suppression of the SXR-overexpressing HuH7 cells. SXR was stably expressed in the hepatocellular carcinoma cell line, HuH7 (HuH7-SXR cells). (A) HuH7-SXR cells and the negative control, HuH-vec cells, were seeded at a density of 1,000 cells/well and cultured in DMEM with 10% FCS. Cell growth was assayed using the WST-8 tetrazolium salt. Bar represents the mean ± SEM values of relative absorbance at 450 nm for each clone on each day normalized to values at day 0 (n = 3). (B) HuH7-SXR and HuH7-vec cells were seeded at a density of 1,000 cells/well and cultured in phenol red-free DMEM with charcoal/dextran-treated FCS (5%) and vitamin K2 (MK-4; 10 μM) or ethanol (Et). Cell growth was assayed using the WST-8 tetrazolium salt. Bar represents the mean ± SEM values of relative absorbance at 450 nm for each clone on day 4 normalized to values at day 0 (n = 4). N.S., not significant.

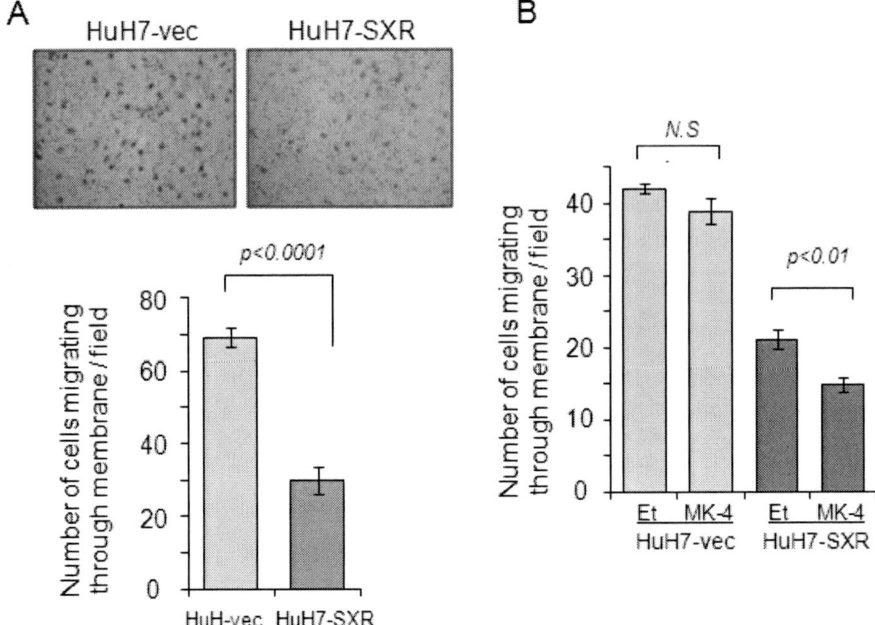

Figure 3. Vitamin K-dependent suppression of motility of the SXR-overexpressing HuH7 cells. (A) HuH7-SXR and HuH-vec cells were seeded on a PET filter (pore size, 8 μm), and their migration activity was evaluated by culturing in DMEM with 10% FCS. The number of cells migrating through the PET filter in 24 h was counted. Upper panels: cells on the lower side of the filters were stained with Giemsa staining solution and visualized under a microscope. Representative views used for cell counting are shown (200× magnification). Lower graph: bars represent the mean number of cells ± SEM counted across 5 fields. (B) Ligand-dependent suppression of motility of the SXR-overexpressing HuH7 cells. HuH7-SXR and HuH7-vec cells were seeded on a PET filter (pore size, 8 μm), and their migration activity was evaluated in phenol red-free DMEM with charcoal/dextran-treated FCS (5%) and vitamin K2 (MK-4; 10 μM) or ethanol (Et). Bars represent the mean number of cells ± SEM counted across 5 fields. N.S., not significant.

The gene encoding the conjugating enzyme, CYP3A4, and that encoding the transporter MDR1 [23] are typical target genes induced by SXR/PXR. SXR/PXR expression was also detected in the kidney, lung [24], bone (especially in osteoblasts) [25], and peripheral mononuclear cells [26].

According to some results from recent investigations, the activation of SXR/PXR in bone tissue could be involved in the favorable effects of vitamin K on bone tissue. Vitamin K induces some genes in osteoblastic cells in an SXR-dependent manner. These genes include tsukushi (*Tsk*), matrilin-2 (*Matn2*), *CD14* [27] and *Msx2* [28]. Among these, *Tsk* encodes the protein having a collagen-accumulating effect [27]. *CD14* affects osteoblastogenesis and osteoclastogenesis, and *Msx2* induces osteoblastogenesis [28]. The collagen-accumulating effect of *Tsk* was not suppressed by warfarin, indicating that this mechanism is not involved in the activation of GGCX [27]. PXR-knockout mice exhibited low bone mineral density with suppressed bone formation and activated bone resorption [29]. These results support the PXR-dependent effects on bone metabolism. On the other hand, vitamin K is a promising agent for cancer treatment. An epidemiological study in North America revealed that postmenopausal women treated with vitamin K1 had a lower risk of cancer than those who received placebo [9]. In this study, almost half of the detected malignancies were breast cancers. The administration of vitamin K2 reduced the de novo occurrence of hepatocellular carcinoma arising from liver cirrhosis in a clinical trial [6]. It was also revealed that the administration of vitamin K2 prolonged the time interval to hepatocellular carcinoma recurrence after treatment [7, 8].

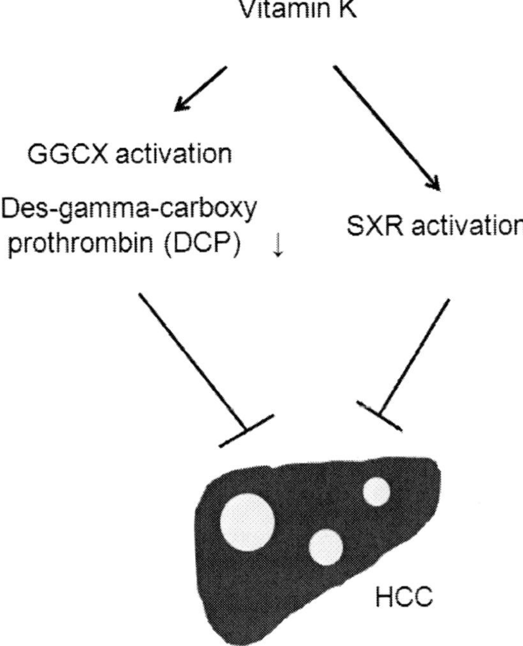

Figure 4. Dual actions of vitamin K on hepatocellular carcinoma. The antitumor effect of vitamin K on hepatocellular carcinoma (HCC) can be attributed to 2 distinct mechanisms: the conventional mechanism of GGCX activation, and the novel mechanism of SXR activation. Existence of these 2 mechanisms can be presumed in other tissues showing simultaneous GGCX and SXR expression.

One of the explanations for the antihepatocellular carcinoma effect of vitamin K is the action of des-γ-carboxyprothrombin (DCP; also known as PIVKA II), which is produced under vitamin K-deficient conditions and has a tumor-promoting effect [30].

Because SXR/PXR is abundantly expressed in the liver, we assumed that the tumor-suppressing effect of vitamin K can be attributed to the SXR/PXR-dependent effect of vitamin K at least in part. Therfore, we utilized hepatocellular carcinoma cell lines, HepG2 and HuH7, that express endogenous SXR.

Upon SXR overexpression, the HuH7 cells, which normally have lower amounts of endogenous SXR than HepG2 cells, displayed suppressed proliferation and motility. These effects of SXR on cell proliferation and motility were further evident in the presence of vitamin K (Figure 2, 3), indicating the existence of the SXR-dependent vitamin K effects [31].

Vitamin K is a natural and safe nutrient. The novel effect of vitamin K that involves the nuclear receptor SXR/PXR has been elucidated. It can be presumed that this novel effect exists in other organs and tissues where SXR/PXR expression is detected, like in the bone (Figure 1) and in the liver (Figure 4).

Further investigation on this mechanism is warranted to understand the precise function of vitamin K and the pathology of the concerning diseases, and to discover new therapeutic approaches for these diseases.

REFERENCES

[1] Okano T, Shimomura Y, Yamane M, Suhara Y, Kamao M, Sugiura M, et al. Conversion of phylloquinone (Vitamin K1) into menaquinone-4 (Vitamin K2) in mice: two possible routes for menaquinone-4 accumulation in cerebra of mice. *J. Biol. Chem.* 2008;283(17):11270-9.

[2] Shiraki M, Shiraki Y, Aoki C, Miura M. Vitamin K2 (menatetrenone) effectively prevents fractures and sustains lumbar bone mineral density in osteoporosis. *J. Bone Miner. Res.* 2000;15(3):515-21.

[3] Cockayne S, Adamson J, Lanham-New S, Shearer MJ, Gilbody S, Torgerson DJ. Vitamin K and the prevention of fractures: systematic review and meta-analysis of randomized controlled trials. *Arch. Intern. Med.* 2006;166(12):1256-61.

[4] Neogi T, Booth SL, Zhang YQ, Jacques PF, Terkeltaub R, Aliabadi P, et al. Low vitamin K status is associated with osteoarthritis in the hand and knee. *Arthritis Rheum.* 2006;54(4):1255-61.

[5] Geleijnse JM, Vermeer C, Grobbee DE, Schurgers LJ, Knapen MH, van der Meer IM, et al. Dietary intake of menaquinone is associated with a reduced risk of coronary heart disease: the Rotterdam Study. *J. Nutr*. 2004;134(11):3100-5.

[6] Habu D, Shiomi S, Tamori A, Takeda T, Tanaka T, Kubo S, et al. Role of vitamin K2 in the development of hepatocellular carcinoma in women with viral cirrhosis of the liver. *JAMA*. 2004;292(3):358-61.

[7] Mizuta T, Ozaki I, Eguchi Y, Yasutake T, Kawazoe S, Fujimoto K, et al. The effect of menatetrenone, a vitamin K2 analog, on disease recurrence and survival in patients with hepatocellular carcinoma after curative treatment: a pilot study. *Cancer*. 2006;106(4):867-72.

[8] Kakizaki S, Sohara N, Sato K, Suzuki H, Yanagisawa M, Nakajima H, et al. Preventive effects of vitamin K on recurrent disease in patients with hepatocellular carcinoma arising from hepatitis C viral infection. *J. Gastroenterol. Hepatol*. 2007;22(4):518-22.

[9] Cheung AM, Tile L, Lee Y, Tomlinson G, Hawker G, Scher J, et al. Vitamin K supplementation in postmenopausal women with osteopenia (ECKO trial): a randomized controlled trial. *PLoS Med*. 2008;5(10):e196.

[10] Varnum BC, Young C, Elliott G, Garcia A, Bartley TD, Fridell YW, et al. Axl receptor tyrosine kinase stimulated by the vitamin K-dependent protein encoded by growth-arrest-specific gene 6. *Nature*. 1995;373(6515):623-6.

[11] Price PA, Otsuka AA, Poser JW, Kristaponis J, Raman N. Characterization of a gamma-carboxyglutamic acid-containing protein from bone. *Proc. Natl. Acad. Sci. U S A*. 1976;73(5):1447-51.

[12] Luo G, Ducy P, McKee MD, Pinero GJ, Loyer E, Behringer RR, et al. Spontaneous calcification of arteries and cartilage in mice lacking matrix GLA protein. *Nature*. 1997;386(6620):78-81.

[13] Coutu DL, Wu JH, Monette A, Rivard GE, Blostein MD, Galipeau J. Periostin, a member of a novel family of vitamin K-dependent proteins, is expressed by mesenchymal stromal cells. *J. Biol. Chem*. 2008;283(26): 17991-8001.

[14] Kulman JD, Harris JE, Haldeman BA, Davie EW. Primary structure and tissue distribution of two novel proline-rich gamma-carboxyglutamic acid proteins. *Proc. Natl. Acad. Sci. U S A*. 1997;94(17):9058-62.

[15] Kulman JD, Harris JE, Xie L, Davie EW. Identification of two novel transmembrane gamma-carboxyglutamic acid proteins expressed broadly in fetal and adult tissues. *Proc. Natl. Acad. Sci. U S A*. 2001;98(4):1370-5.

[16] Kinoshita H, Nakagawa K, Narusawa K, Goseki-Sone M, Fukushi-Irie M, Mizoi L, et al. A functional single nucleotide polymorphism in the vitamin-

K-dependent gamma-glutamyl carboxylase gene (Arg325Gln) is associated with bone mineral density in elderly Japanese women. *Bone.* 2007;40(2):451-6.

[17] Vergnaud P, Garnero P, Meunier PJ, Bréart G, Kamihagi K, Delmas PD. Undercarboxylated osteocalcin measured with a specific immunoassay predicts hip fracture in elderly women: the EPIDOS Study. *J. Clin. Endocrinol. Metab.* 1997 Mar;82(3):719-24.

[18] Ducy P, Desbois C, Boyce B, Pinero G, Story B, Dunstan C, et al. Increased bone formation in osteocalcin-deficient mice. *Nature.* 1996;382(6590):448-52.

[19] Zhou C, Verma S, Blumberg B. The steroid and xenobiotic receptor (SXR), beyond xenobiotic metabolism. *Nucl. Recept. Signal.* 2009;7:e001

[20] Staudinger JL, Goodwin B, Jones SA, Hawkins-Brown D, MacKenzie KI, LaTour A, et al. The nuclear receptor PXR is a lithocholic acid sensor that protects against liver toxicity. *Proc. Natl. Acad. Sci. U S A.* 2001;98(6):3369-74.

[21] Xie W, Radominska-Pandya A, Shi Y, Simon CM, Nelson MC, Ong ES, et al. An essential role for nuclear receptors SXR/PXR in detoxification of cholestatic bile acids. *Proc. Natl. Acad. Sci. U S A.* 2001;98(6):3375-80.

[22] Blumberg B, Sabbagh W Jr, Juguilon H, Bolado J Jr, van Meter CM, Ong ES, et al. SXR, a novel steroid and xenobiotic-sensing nuclear receptor. *Genes Dev.* 1998;12(20):3195-205.

[23] Synold TW, Dussault I, Forman BM. The orphan nuclear receptor SXR coordinately regulates drug metabolism and efflux. *Nat. Med.* 2001;7(5):584-90.

[24] Miki Y, Suzuki T, Tazawa C, Blumberg B, Sasano H. Steroid and xenobiotic receptor (SXR), cytochrome P450 3A4 and multidrug resistance gene 1 in human adult and fetal tissues. *Mol. Cell Endocrinol.* 2005;231(1-2):75-85.

[25] Tabb MM, Sun A, Zhou C, Grün F, Errandi J, Romero K, et al. Vitamin K2 regulation of bone homeostasis is mediated by the steroid and xenobiotic receptor SXR. *J. Biol. Chem.* 2003;278(45):43919-27.

[26] Albermann N, Schmitz-Winnenthal FH, Z'graggen K, Volk C, Hoffmann MM, Haefeli WE, et al. Expression of the drug transporters MDR1/ABCB1, MRP1/ABCC1, MRP2/ABCC2, BCRP/ABCG2, and PXR in peripheral blood mononuclear cells and their relationship with the expression in intestine and liver. *Biochem. Pharmacol.* 2005;70(6):949-58.

[27] Ichikawa T, Horie-Inoue K, Ikeda K, Blumberg B, Inoue S. Steroid and xenobiotic receptor SXR mediates vitamin K2-activated transcription of

extracellular matrix-related genes and collagen accumulation in osteoblastic cells. *J. Biol. Chem.* 2006;281(25):16927-34.

[28] Igarashi M, Yogiashi Y, Mihara M, Takada I, Kitagawa H, Kato S. Vitamin K induces osteoblast differentiation through pregnane X receptor-mediated transcriptional control of the Msx2 gene. *Mol. Cell Biol.* 2007;27(22):7947-54.

[29] Azuma K, Casey SC, Ito M, Urano T, Horie K, Ouchi Y, et al. Pregnane X receptor knockout mice display osteopenia with reduced bone formation and enhanced bone resorption. *J. Endocrinol.* 2010:207(3):257-63

[30] Ma M, Qu XJ, Mu GY, Chen MH, Cheng YN, Kokudo N, et al. Vitamin K2 inhibits the growth of hepatocellular carcinoma via decrease of des-gamma-carboxy prothrombin. *Chemotherapy.* 2009;55(1):28-35.

[31] Azuma K, Urano T, Ouchi Y, Inoue S. Vitamin K2 suppresses proliferation and motility of hepatocellular carcinoma cells by activating steroid and xenobiotic receptor. *Endocr. J.* 2009;56(7):843-9.

In: Nuclear Receptors
Editors: M. K. Bates and R. M. Kerr
ISBN: 978-1-61209-980-4

Chapter 6

ANDROGEN RECEPTORS HAVE A POTENTIAL ROLE IN MEDIATING THE SEROTONIN SYNTHESIS MECHANISM

Takahiro Fukumoto[1,*], Noriko Tosa[2] and Tadaaki Miyazaki[3]

[1]Research Center for Infection-associated Cancer, Institute for Genetic Medicine, Hokkaido, University, Kita-15, Nishi-7, Kita-ku, Sapporo, 060-0815,Japan

[2]Institute for Animal Experimentation, Hokkaido University Graduate School of Medicine, Kita-15, Nishi-7, Kita-ku, Sapporo, Hokkaido, 060-8638, Japan, Hokkaido University, Kita-15, Nishi-7, Kita-ku, Sapporo, 060-0815,Japan

[3]Department of Bioresource, Hokkaidou University Research Center for Zoonosis Control, North 20, West 10, Kita-ku, Sapporo, Japan, 0010020

ABSTRACT

Nuclear receptors are a class of proteins that have the ability to directly bind to DNA and regulate gene expression, and these receptors are classified as transcription factors. This report focuses on a new function of AR (androgen receptor). Androgen receptors (ARs) belong to the steroid receptor family and play an essential role in the generation and development of the prostate. Androgen receptors have similar conserved domains that are

* Tel. and Fax: +81-11-706-6053, E-mail: takataka@igm.hokudai.ac.jp

composed of an NTD (N-terminal domain), a DBD (DNA-binding domain), and an LBD (ligand-binding domain). The NTD works stabilize bound androgen and the AR-LBD mediates the interaction between AR and other proteins, which include Hsps (heat-shock proteins). In the absence of androgen, AR remains in the cytoplasm in an inactive form. After AR binds to androgens, activated AR can bind with other signal molecules and form functional complexes. Then, the complex translocates to the nucleus and regulates the gene expression for androgen regulated genes. Recently, some research has shown that AR can interact with DDC (L-dopa decarboxylase), a key molecule for serotonin (5-HT) synthesizing. Serotonin is a well known neurotransmitter but has been mentioned in the relationship with the generation of the prostate. Then, we introduce here that AR can regulate prostate cancer progression via the serotonin synthesis process.

A suggested rewrite of the previous sentence, placed here to avoid ambiguity: This paper suggests that AR may play a role in regulating the progress of prostate cancer via the serotonin synthesis process.

INTRODUCTION

Normal and neoplastic growth of tissue, especially the prostate gland and the testis, depend on the androgen receptor (AR) expression and function. The mechanisms by which AR exerts this control, however, remains poorly understood. Here, we will present an overview of new evidence for understanding the contribution of the molecular mechanism on the contribution of the androgen receptor.

Androgen Receptors Has Several Roles for Our Life

The androgen receptor (AR) is a type of nuclear receptor, which is activated by binding with of either of the androgenic hormones testerone or dihydrotestosterone in the cytoplasm and then translocating into the nucleus [1]. The main function of the androgen receptor is as a DNA binding transcription factor, which regulates gene expression, and one of the known target genes of androgen receptor activation is insulin-like growth factor I (IGF-1)[2]. In humans, androgen receptors are encoded by the AR gene located on the X chromosome at Xq11-12 [3] and seems to be involved in neuron physiology [4] and disturbances in fertility [5]. In addition, point mutations and trinucleotide repeat polymorphism has been linked to a number of additional disorders including prostate cancer [6]. It is also known that there are several transgenic mice models with an ubiquitous knockout of the AR (ARKO). The first animal model of AR was described in 1970 by Lyon and Hawkes [7] and nineteen further transgenic mice were already reported. Most phenotypes indicate both a female-like external appearance, testes are reduced in size and located intra-abdominally, and also generation tumorization [8]. Therefore, AR plays an essential role in the development of the life of organisms and it is important to know details of the functions involved.

Androgen Receptors Can Bind Directly with Dopa Decarobxylase

We have learnt that AR can bind with other protein in the cytoplasm to cause signaling by androgens. (Refer the web site for molecules interacting with AR.: The Androgen receptor gene mutations database world wide web server (http://androgendb.mcgill.ca/ARinteract.pdf)]

There are two major pathways proposed [8] and the potential roles in several tissue types will be discussed separately next. This chapter focuses on evidence that AR can bind with dopa decarboxylase (DDC), a serotonin synthesis enzyme, as reported by Paul S. Rennie and coworkers in 2003 [9]. They identified DDC as a co-factor for AR activation by using a yeast two-hybrid system and proposed a mechanism during generation of prostate cancer. This was the first publication suggesting that serotonin synthesis has a potential role in prostate cancer progression with AR regulation although serotonin signaling via its receptors have been know to have an effect on prostate cancer [10]. According to this publication, DDC can bind with the N-terminal domain of AR in vitro and in vivo. Further, the transient expression of DDC in prostate cancer cell lines strongly enhanced ligand-dependent AR transcriptional activity and this publication [10] showed that the interaction DDC and AR have important implications in prostate cancer progress.

Dopa Carboxylase Synthesizes Serotonin and Has a Role in the Progress of Cancer

Dopa decarboxylase (DDC) is a pyridoxal 5'-phosphate-dependent enzyme that catalyses the decarboxylation of 5-hydroxytryptophan (5-HTP) to serotonin [11]. It is also commonly referred to as aromatic-L-amino-acid decarboxylase (AADC). This enzyme has also been found to decarboxylate other aromatic L-amino acids. Further, DDC has been purified from several sources [12] although it has been thought to be a rather unregulated molecule. Serotonin is a well known neurotransmitter and it has been suggested that DDC activity regulation could be of importance during treatment of neural diseases including Parkinson's disease. Serotonin has been implicated in the relationship with tumors and several antagonists for serotonin receptors have been reported to actually inhibit tumor progression [10]. Recently, it has been demonstrated that in small cell lung carcinomas and in neuroblastoma, neoplasias with increased DDC activity and some other types of tumors are also characterized by an increase in DDC activity [13]. These observations have lead researchers to suggest that the DDC could be a possible molecular target for the action of enzyme-activation in the cytoplasm. Paul S. Rennie has demonstrated that DDC interacts with AR and regulates prostate cancer progress. We have suggested that additional serotonin could induce the growth of prostate cancer cell lines, and a specific knockdown of DDC using siRNA indicated the complete inhibition of tumor cell development [14].

It must be borne in mind that synthesis of serotonin via DDC activity is essential for cell growth in both non-cancerous (normal) and cancer cells. Also of note, we recently identified the cellular adhesion molecule family which binds with the AR-DDC complex (submitted) and this molecule can regulate transcriptional activity dependent on AR (unpublished data). These experimental findings would allow the postulation of a new hypothesis for the function of AR and provide new insight into AR as a nuclear receptor. The next section here introduces a classical hypothesis, which may be explained by invoking the functions of AR introduced above.

AR May Be a Key Molecule for Explaining the APUD Concept

Now we will here introduce a classical concept, termed the APUD concept, which may contribute to understanding the molecular mechanism of the AR function. The APUD is the acronym for amine and precursor uptake and decarboxylation, and it was proposed as a concept for understanding human bodily homeostasis focusing on the variety of diffusely scattered endocrine cells by Dr. Pearse in 1968 [15]. This concept proposes that quite a number cells can induce 'internal secretion of amine' in every tissue of animals and that these cells regulate the basis of life (and diseases) by using the amines as chemical messengers. Here it is of importance that the cells are not only capable of amine processing and production of polypeptide hormones but also metabolize via the endocrine and autocrine system. Thus, these cells are not only coordinated with each other in respect to the production of peptides and amines with hormonal activity (as well as paracrine hormones and neurotransmitters) but are also coupled with the autonomic and somatic system as super-ordinate controllers. According to the concept, we wish to suggest that AR has a potential role in regulating our life stages. There are a number of reasons, 1) AR is present in many kinds of tissue, 2) AR is present in the cytoplasm of cells and can bind with DDC (in the time, we can detect serotonin), 3) AR binding with DDC translocates to the nucleus (in the stage where we cannot detect serotonin), 4) AR transcription activity enhances MAOA gene expression (Figure 1). Therefore, AR can regulate the release (as well as synthesis and metabolism) of serotonin via the interaction with DDC and may coordinate body homeostasis. If the process is blocked by some unknown mechanism, which means that there has been an upset in the AR-DDC regulation balance, 1) serotonin is synthesized in huge amounts and enhance abnormal cell growth (ex, cancer progression [14]), 2) MAOA metabolizes all serotonin and induces neuronal diseases like depression, abnormal organ formations (ex, polycystic ovary syndrome [16,17]), or cell death [18]. There is also a good example in prostate cancer, PC-3 cells, AR-independent prostate cancer cell lines, these express DDC not AR and release serotonin in about 5 times the amounts of an unaffected prostate cell line (PrEC). This cell line does not suffer from effects of AR regulation and continues to release serotonin for growth like in heavy prostate cancer progression [14]. Thus, AR may regulate the APUD cell function with AR transcriptional activity.

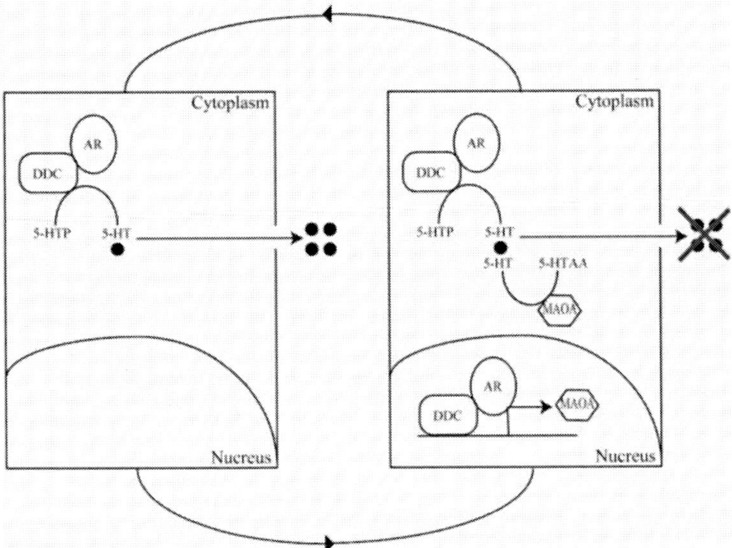

Figure 1. AR regulates the balance of serotonin synthesis and metabolism system.

AR Regulation System for DDC May Coordinate a Sex-Determination in Animal

AR is expressed in the kidney as well as in urological organs but its expression pattern is different between males and females [19]. Similar to AR, DDC also expresses in the kidney and indicates a similar localization ([20], unpublished data). The AR and DDC show high molecular expressions in males and and this would support the phenotypes from ARKO mouse that indicate female-like observations. According to our experiments, the AR-DDC system plays a role in both kidney organogenesis and in sex-determination in a few species. Overall, it seems possible, even likely that brand new functions for AR may be proposed by.

In Conclusion

AR belongs to the nucleus receptor family and is expressed in a variety of kinds of tissue. Its transcriptional activity plays an essential role as a basis for life as well as it has other functions as a co-factor for enzymes or signaling molecules. In this chapter, we have focused on the interaction between AR and DDC and introduced a new hypothesis about AR. AR has multiple functions and may have more critical effects that we do not know yet. Therefore, our adventure for understanding AR functions will continue for a long time from now also.

References

[1] Nuclear receptors enhance our understanding of transcription regulation. *Green S and Chambon* P. TIG.4: 309-314.1988.

[2] Androgens up-regulate the insulin-like growth factor-I receptor in prostate cancer cells. Pandini G, Mineo R, Frasca F, Roberts CT Jr, Marcelli M, Vigneri R and Belfiore A. *Cancer Res*. 65 (5): 1849–1857. 2005.

[3] Molecular cloning of human and rat complementary DNA encoding androgen receptors. Chang CS, Kokontis J and Liao ST. *Science,* 240 (4850): 324–326.1988.

[4] Progressive proximal spinal and bulbar muscular atrophy of late onset. A sex-linked recessive trait. Kennedy WR, Alter M and Sung JH.*Neurology* 18 (7): 671–680.1968.

[5] Androgen receptor defects: historical, clinical, and molecular perspectives. Quigley CA, De Bellis A, Marschke KB, el-Awady MK, Wilson EM and French FS. *Endocr. Rev.* 16(3):271-321.1995.

[6] The androgen receptor gene mutations database (ARDB). Gottlieb B, Lenore K, Beitel LK, Wu JH and Trifiro M. *Human Mutation.* 23(6): 527–533.2004.

[7] X-linked gene for testicular feminization in the mouse. Lyon MF and Hawkes SG. *Nature.* 227 :1217–1219. 1970.

[8] Androgens and spermatogenesis: lessons from transgenic mouse models. Verhoeven G, Willems A, Denolet E, Swinnen JV, De Gendt K. *Philos. Trans. R. Soc. Lond. B. Biol. Sci.* 365(1546):1537-1556.2010.

[9] Isolation and identification of L-dopa decarboxylase as a protein that binds to and enhances transcriptional activity of the androgen receptor using the repressed transactivator yeast two-hybrid system. Wafa LA, Cheng H, Rao MA, Nelson CC, Cox M, Hirst M, Sadowski I and Rennie PS. *Biochem. J.* 375:373-383. 2003.

[10] The role of serotonin in tumour growth (review). Siddiqui EJ, Thompson CS, Mikhailidis DP and Mumtaz FH. *Oncol. Rep.* 14(6):1593-1597.2005.

[11] Serotonin signaling is a very early step in patterning of the left-right axis in chick and frog embryos. Fukumoto T, Kema IP and Levin M. *Curr. Biol.* 15(9):794-803.2005.

[12] Dopa Decarboxylase: A model Gene-Enzyme system for studying development, Behavior, and systematics. Hodgetts RB and O'Keefe SL. *Annu. Rev. Entomol.* 51: 259-284. 2006.

[13] Notch-1 regulates pulmonary neuroendocrine cell differentiation in cell lines and in transgenic mice. Shan L, Aster JC, Sklar J and Sunday ME. *Am. J. Physiol. Lung Cell Mol. Physiol.* 292(2): L500-509. 2007.

[14] Shinka,T. ------ Fukumoto.T 2010 (on the revising).

[15] Common cytochemical and ultrastructural characteristics of cells producing polypeptide hormones (the APUD series) and their relevance to thyroid and ultimobrachial C cells and calcitonin. Pearse AGE. *Proceedings of the Royal Society London B.* 170: 71-80. 1968.

[16] Dynamic Content Exchange of Serotonin in mouse oocyte maturation. Fukumoto T, Shoji N, Onodera D, Sakurai I, Hirano Y, Tanaka T and Hatamura I. *JMOR.* 27: 216-219.2010.

[17] Cytological changes of ovarian oocytes in the polycystic ovary induced by injection of testosterone propionate in mice. Sakurai I, Kouda T, Fukumoto T, Hatamura I and Y. Hirao. *JMOR.* 27: 220-224. 2010.

[18] The impact of cell adhesion changes on proliferation and survival during prostate cancer development and progression. Knudsen BS and Miranti CK. *J. Cell Biochem.* 99(2):345-361.2006.

[19] Sex-specific regulation of ENaC and androgen receptor in female rat kidney.Kienitz T, Allolio B, Strasburger CJ and Quinkler M. *Horm. Metab. Res.* 41(5):356-362. 2009.

[20] Opposite sexual dimorphism of 3,4-dihydroxyphenylalanine decarboxylase in the kidney and small intestine of mice. López-Contreras AJ, Galindo JD, López-García C, Castells MT, Cremades A and Peñafiel R. *J. Endocrinol.* 196(3):615-624.2008.

In: Nuclear Receptors
Editors: M. K. Bates and R. M. Kerr
ISBN: 978-1-61209-980-4

Chapter 7

THE EXPORTIN1 GENES (XPO1A AND XPO1B) IN ARABIDOPSIS: ARE THEY FUNCTIONALLY REDUNDANT?

Lian-Chin Wang [*] ***and Shaw-Jye Wu***
Department of Life Sciences, National Central University, No.300 Jhong-da Road, Jhong-li City, Taoyuan 32001, Taiwan

Plants are immobile and, therefore, confronted with a variety of environmental stresses. To adapt to these harmful conditions, plants have developed effective and complex stress signal transduction pathways between the nucleus and cytoplasm for response to and survival from stress conditions. In eukaryotic cells, the nuclear envelope separates the cytoplasm from the nucleus, and the nuclear pore complex (NPC) is the gateway for signal molecules trafficking across the nuclear envelope (Meier and Brkljacic, 2009). Small molecules utilize passive diffusion to pass through the NPC; however, the efficient and directed translocation of macromolecules requires nuclear transport receptors to facilitate the passage through the NPC. In this regard, importinβ-like nuclear transport receptors are the main receptors for nuclear transport in Arabidopsis. Some of these receptors act as nuclear import receptors (importins) and some as nuclear export receptors (exportins) (Merkle, 2004; Merkle, 2008). In addition, many stress responses in plants are controlled by the nucleocytoplasmic partitioning of regulatory molecules between the cytoplasm and nucleoplasm,

[*] Tel: +886-3-422-7971, Fax: +886-3-422-8482, E-mail: dcmotoya@yahoo.com.tw

such as light, temperature, and responses to cytokinin-signal transduction and pathogen infection (Meier, 2005). Thus, importins and exportins are very important for plant growth, development, and stress response.

In Arabidopsis, at least 18 importinβ-like nuclear transport receptors have been identified (Huang et al., 2010); nevertheless, only few of these receptors have been functionally characterized. The receptors that have been well characterized include PAUSED (PSD), which was discovered to export tRNA (Hunter et al., 2003), and HASTY (HST), which was shown to be involved in the export of smRNA to the cytoplasm (Bollman et al., 2003). These 2 importinβ-like nuclear transport receptors play a relatively important role, as they are responsible for tRNA and smRNA nuclear export. However, loss-of-function mutations in PSD and HST have demonstrated a pleiotropic but non-lethal phenotype; thus, Arabidopsis may have other genes that partially complement the functions of PSD and HST. EXPORTIN 1 (XPO1) is another characterized exportin, which transports leucine-rich nuclear export signal (NES)-containing proteins. Unlike yeast, humans, *Drosophila*, and *Caenorhabditis elegans*, which have only 1 *XPO1* gene (von Koskull-Döring et al., 2007); however, there are 2 copies of the *XPO1* gene (*XPO1A* and *XPO1B*) in Arabidopsis. According to the alignment results of the amino acid sequence, gene structure, and protein structure, XPO1A and XPO1B share a high identity (Huang et al., 2010). A previous study showed that Arabidopsis with a single mutation in either *XPO1A* or *XPO1B* had a normal appearance, but a double mutant of these 2 genes was gametophyte-defective (Blanvillain et al., 2008), which implies that XPO1A and XPO1B may be functionally redundant. However, in our recent study using a forward genetic approach, we screened a *heat-intolerant 2* (*hit2*) Arabidopsis mutant and mapped the mutated gene to the XPO1A gene (Wu et al., 2010). The hypersensitive phenotype to basal thermotolerance in *XPO1A* did not display in *XPO1B*, indicating that XPO1A has a different function from XPO1B under heat stress conditions.

Despite these results, it remains unclear whether the functions of *XPO1A* and *XPO1B* are interchangeable in other tissue types, growth stages, or stress conditions and whether they have a specific function. Further, it is still unclear how to distinguish the substrates of these 2 genes. Protein sequence alignment has shown that XPO1A and XPO1B have 86% identity; moreover, these paralogs were considered to originate from the same ancestors by gene duplication (Hung et al., 2010). According to the information from the Arabidopsis database (http://bar.utoronto.ca/efp/cgi-bin/efpWeb.cgi), the transcript levels of *XPO1A* in various tissues are higher than those of *XPO1B,* and only in mature pollen is the content of *XPO1A* lower than that of *XPO1B*. A similar phenomenon was also found for other importinβ-like transport receptor genes, for example, *AtIMPα3* (AT4G02150)-*AtIMPα6* (AT1G02690) and *AtIMPβ6* (AT2G31660)-*AtIMPβ12* (AT3G59020), which are also considered to be duplicated genes. The expression level of *AtIMPα3* and *AtIMPβ12* in different tissues has been shown to be higher than that of *AtIMPα6* and *AtIMPβ6* (Hung et al., 2010). In addition, *AtIMPβ6* exhibited a specific function in abscisic acid signal transduction that was different from that of *AtIMPβ12*. Likewise, XPO1A not only has a different function from XPO1B under long-term heat stress and heat shock treatment (Wu et al., 2010), but XPO1A is also more sensitive to salt stress (unpublished result). These results imply a functional difference between XPO1A and XPO1B in different tissues, growth stages, and stress conditions.

These differences between XPO1A and XPO1B may be due to the fact that they may have specific adaptors for their distinct substrates and that they may recognize these particular substrates by phosphorylation or dephosphorylation *of* the substrates (Hutten and Kehlenbach, 2007) or by utilizing different small GTPase Ras-related nuclear proteins (Ran) to distinguish the substrates. It is well known that XPO1 requires the interaction with small GTPase Ran in its GTP-bound form (RanGTP) to enable nuclear export. RanGTPase is crucial for triggering the formation of the export complex and to release the export substrate into the cytoplasm (Lui and Huang, 2009). The absence of RanGTPase has been shown to lead to XPO1 with a low affinity for the substrate. In Arabidopsis, 4 RanGTPs have been identified, namely, *AtRAN1*, *AtRAN2*, and *AtRAN3*, which have 95%–96% identity and an identical conserved motif, and *AtRAN4*, which differs from other AtRANs and is a salt stress-inducible protein (Vernoud et al., 2003). As mentioned above, our unpublished data also indicated that XPO1A, but not XPO1B, is more sensitive to salt stress. These results imply that XPO1s may function either to recognize the specific substrate or to interact with specific RanGTPs. On the other hand, *XPO1A* and *XPO1B* may be able to recognize and transport the same cargos, but they may have a different affinity to do so or may use different adaptor proteins to interact with identical substrates. Thus, many features of these receptors may yet be complementary. Therefore, with only a single gene mutation, the other gene may exhibit poor substrate binding; hence Arabidopsis is still able to function normally under normal conditions.

Owing to the fact that plants cannot flee from harmful conditions, they have evolved many stress signals, such as heat shock transcription factors (Hsfs), for fast and efficient stress responses to ensure survival. Hsfs vary among eukaryotic organisms, with *Saccharomyces*, *Drosophila*, and *C. elegans* having 1 Hsf, whereas vertebrates have 4 Hsfs. Plants have a fairly complex Hsf family, for example, Arabidopsis and rice have 21 and 25 Hsfs, respectively (Guo et al.,

2008; Nover et al., 2001). Most Hsfs contain nuclear localization signals (NLS) and NES, which can change their intracellular distribution between the nucleus and cytoplasm in response to heat stress (Guo et al., 2008). Further, as mentioned above, XPO1 has been shown to interact mainly with NES-containing proteins, which implies that XPO1 is crucial for the plant heat stress response at the control level of nucleo-cytoplasmic partitioning of Hsfs. However, due to the large number and highly complex network of Hsfs, it appears impossible for only 1 exportin to be responsible for the transport of so many stress signals and other NES-containing proteins. Hence, the 2 *XPO1* genes in Arabidopsis probably originated from the neofunction of adapting to varying environments.

The relationship between XPO1A and XPO1B appears to be very complicated; however, the differences between XPO1A and XPO1B described here provide new insights into the nucleo-cytoplasmic trafficking involved in the plant stress response. The *hit2* mutant revealed that Arabidopsis has evolved a stress tolerance-specific nuclear export receptor (XPO1A) that is specifically required for the stress response, and XPO1A inevitably has specific substrates that require nucleo-cytoplasmic partitioning under stress conditions. Nevertheless, even though preliminary results indicate some differences between XPO1A and XPO1B, several fundamental questions remain unanswered. For example, what is the difference in function between XPO1A and XPO1B, and how does one distinguish the interaction of XPO1A and XPO1B with a large number of substrates? Further, it is imperative that the specific substrates for XPO1A and XPO1B be identified. Many plant-specific features of nuclear transport mechanisms are not well characterized; further study will provide valuable information about the nucleo-cytoplasmic trafficking by XPO1s in plants.

References

Blanvillain R., Boavida L. C., McCormick S., Ow D. W. (2008) *EXPORTIN1* Genes Are Essential for Development and Function of the Gametophytes in Arabidopsis thaliana. *Genetics* 180:1493-1500.

Bollman K. M., Aukerman M. J., Park M. Y., Hunter C., Berardini T. Z., Poethig R. S. (2003) HASTY, the Arabidopsis ortholog of exportin 5/MSN5, regulates phase change and morphogenesis. *Development* 130:1493–1504.

Guo J., Wu J., Ji Q., Wang C., Luo L., Yuan Y., Wang Y., Wang J. (2008) Genome-wide analysis of heat shock transcription factor families in rice and Arabidopsis. *Journal of Genetics and Genomics,* 35:105−118.

Huang J. G., Yang M., Liu P., Yang G. D., Wu C. A., Zhen C. C. (2010) Genome-wide profiling of developmental, hormonal or environmental responsiveness of the nucleocytoplasmic transport receptors in Arabidopsis. *Gene* 451: 38-44.

Hunter C., Aukerman M. J., Sun H., Fokina M., Poethig R. S. (2003) PAUSED encodes the Arabidopsis orthologue of exportin-t. *Plant Physiol.* 132:2135–2143.

Hutten S. and Kehlenbach R. H. (2007) CRM1-mendated nuclear export: to the pore and beyond. *TRENDS in Cell Biology*, 17: 193-201.

Lui K. and Huang Y. (2009) RanGTPase: A Key Regulator of Nucleocytoplasmic Trafficking. *Mol. Cell Pharmacol.* 1: 148–156.

Meier, I. (2005) Nucleocytoplasmic trafficking in plant cells. *Int. Rev. Cytol.* 244: 95-135.

Merir I. and Brkljacic J. (2009) The nuclear pore and plant development. *Curr. Opin. Plant Biol.* 12:87-95.

Merkle T. (2004) Nucleo-cytoplasmic partitioning of proteins in plants: implications for the regulation of environmental and developmental signaling. *Curr. Genet.* 44: 231–260.

Merkle T. (2008) Nuclear export of proteins and RNA. *Springer Series Plant Cell Monogr.* 14: 55-77.

Nover L., Bharti K., Doring P., Mishra S. K., Ganguli A., Scharf K. D. (2001) Arabidopsis and the heat stress transcription factor world: how many heat stress transcription factors do we need? *Cell Stress Chaperones,* 6: 177–189.

Vernoud V., Horton A. C., Yang Z., Nielsen E. (2003) Analysis of the small GTPase gene superfamily of Arabdiopsis. *Plant Physiol.* 131: 1191-1208.

von Koskull-Döring P., Scharf K. D., Nover L. (2007) The diversity of plant heat stress transcription factors. *Trends Plant Sci.* 12:452-457.

Wu S. J., Wang L. C., Yeh C. H., Lu C. A., Wu S. J. (2010) Isolation and characterization of the Arabidopsis *hit-intolerant 2 (hit2*) mutant reveal the essential role of the nuclear export receptor EXPORTIN1A (XOP1A) in plant heat tolerance. *New Phytologist.* 186: 833-842.

In: Nuclear Receptors
Editors: M. K. Bates and R. M. Kerr

ISBN: 978-1-61209-980-4

Chapter 8

Peroxisome Proliferator-Activated Receptors: Nuclear Receptors with Pleotropic Actions

Nik Soriani Yaacob[1,*] and Mohd Nor Norazmi[2]
[1]Department of Chemical Pathology, School of Medical Sciences and
[2]School of Health Sciences, Universiti Sains Malaysia, Health Campus, Kubang Kerian, Kelantan, Malaysia

Abstract

Peroxisome proliferator-activated receptors (PPARs) are ligand-activated nuclear receptors that regulate gene expression and are modulated by interaction with corepressors and coactivators. Natural and synthetic ligands promote heterodimerization of PPARs with the retinoid-X-receptor (RXR), facilitating their binding to consensus DNA sequences on target genes. To date, three subtypes of PPAR have been identified – α, δ, and γ; with the γ subtype consisting of two distinct functional isoforms, γ1 and γ2. Although structurally similar, the PPAR subtypes have specific tissue distribution and functions. PPARs regulate multiple cellular functions, such as cell proliferation, the immune response and lipid metabolism and therefore their ligands have been investigated for their potential use in various clinical settings. For example, the PPARγ ligands comprising the thiazolidinedione class of drugs have been used for the management of type 2diabetes. The

* corresponding author (Tel: 609-7676480; Fax: 609-7653370)

potential use of PPAR ligands in autoimmune diseases is also being investigated whereas the specific role of PPARs in tumor development is still controversial. Despite some drawbacks, PPARs still remain as potential therapeutic targets for various conditions and currently dual and pan agonists are being investigated for this purpose. Taken together, the role of PPARs in various cellular processes and disease pathogenesis still requires further investigation and continues to be an exciting field of research. This review will attempt to provide examples of some of the recent findings in these areas of research, highlighting the mechanisms of action and the potential use of PPAR agonists as well as the challenges that still need to be addressed.

INTRODUCTION

The peroxisome proliferator-activated receptor (PPAR) owes its name to its ability to bind the fibrate class of drugs that induce proliferation of peroxisomes in rat hepatocytes (Isseman & Green, 1990). This original PPAR was later re-named PPARα when two other PPAR subtypes, PPARγ and PPARδ, were described. Although both PPARγ and PPARδ do not induce proliferation of peroxisomes, the name is retained since there is high structural homology with PPARα. Since their discovery, the role of PPARs, including that of PPARα, has expanded greatly, and they are now implicated in a variety of physiological and pathological processes. As such, there have been much interest in understanding the cellular pathways modulated by PPARs since aberrations of these pathways may promote the development and progression of diseases such as diabetes, cardiovascular diseases, autoimmune conditions, neurodegenerative diseases and cancer (Michalik et al., 2006; Heneka and Landreth, 2007; Bensinger and Tontonoz, 2008). As transcription factors, PPARs regulate the expression of many genes involved in key cellular regulatory processes and could serve as potential pharmaceutical drug targets for the treatment of many of these diseases. This review aims to briefly discuss the control of PPAR activation and how they can potentially be modulated in selected diseases such as in cardiovascular disease, diabetes, immune regulation and cell proliferation; and to highlight the various challenges in using their agonists for such purposes. This review is not intended to provide an exhaustive description of all the potential roles of PPARs in various diseases and readers are encouraged to refer to the many reviews cited here and elsewhere, on the specific roles of PPARs in these diseases.

PEROXISOME PROLIFERATOR-ACTIVATED RECEPTORS (PPARS)

Peroxisome proliferators are structurally diverse compounds and are generally non-genotoxic (non-mutagenic) in that they do not interact with or damage DNA directly or following metabolic activation (Warren *et al.*, 1980) but they regulate genes encoding key proteins in cellular functions and produce rapid and coordinated pleotropic effects. The receptor that mediate these actions was isolated and characterised from mouse liver by Issemann and Green in 1990 and given the name, peroxisome proliferator-activated receptor (PPAR). Since then, many PPARs have been isolated from other species including humans (Dreyer et al., 1992; Kliewer et al., 1994; Greene et al., 1995; Hotta et al., 1998; Houseknecht et al., 1998). The discovery of PPARs has accelerated the understanding of the intracellular actions of these transcription factors although peroxisome proliferation *per se* has never been observed in humans following PPAR activation. Being a member of the Type II nuclear receptor family, PPARs are localized in the nucleus unlike Type I nuclear receptors which are cytosolic and translocate to the nucleus upon activation by ligand-binding.

PPARs belong to the nuclear hormone receptor superfamily which consists of a group of ligand-activated DNA transcription factors (Isseman and Green, 1990) that regulate expression of genes important for cellular metabolism (Chawla et al., 2001). There are currently 3 known subtypes of PPARs, namely PPARα (NR1C1), PPARδ (NR1C2; also known as NUCI in humans and FAAR [fatty acid-activated receptor] in rodents) and PPARγ (NR1C3) (Kliewer et al, 1994 and Braissant et al., 1996). These subtypes have very similar molecular structure but are encoded by different genes located on human chromosomes 22, 6 and 3, respectively (Bar-Tana, 2001; Rotondo & Davidson, 2002) and mouse chromosomes 15, 17 and 6, respectively (Yousef & Badr, 2004). These subtypes are also highly homologous between various species – for example, mPPARα has 97% and 91% amino acid homology with rPPARα (Motojima, 1993) and hPPARα (Sher *et al.*, 1993), respectively.

PPARs are organized into five functional domains defined as A, B, C, D and E. The A/B domains are the least conserved as well as the most variable in length between the PPARs (Motojima, 1993). This ligand-independent transactivation domain contains the Activation Function 1 (AF1), which is transcriptionally active in the absence of ligands. The C domain is the DNA-binding domain consisting of two zinc fingers and is highly conserved. The P box in the first zinc finger is responsible for specific recognition of the response element (see below) and the D box in the second zinc finger, characteristically containing only three amino acids in PPARs, is involved in dimerization (Wahli *et al.*, 1995). The D domain contains the hinge region (Weigel, 1996) and is the docking site for co-factors. The E domain is the ligand binding domain which is much larger than the DNA binding domain (Green *et al.*, 1993). It consists of 12 α-helical regions, H1 to H12, and is also a binding site for co-activator proteins (Zhang & Young, 2002; Kota *et al.*, 2005; Lazar, 2005). It also contains the Activating Function 2 (AF-2) which mediates conformational changes through a conserved hydrogen-bonding network to create an interacting surface with co-activator proteins upon binding with a ligand (Cronet *et al.*, 2001; Blanquart *et al.*, 2003). A sixth, F domain, which is normally present in nuclear hormone receptors is absent in PPAR (Green *et al.*, 1992). This domain was originally found in the estrogen receptor (ER) and does not seem to be important for ligand binding (Weigel, 1996).

PPARα is a key regulator of energy homeostasis and is therefore highly expressed in the liver, kidney, intestine, heart and skeletal muscle which are involved in fatty acid oxidation and cholesterol metabolism (Lefebvre et al., 2006). PPARα is also expressed in macrophages, T lymphocytes, vascular endothelial cells and smooth muscle cells (Zhang and Young, 2002). Natutral and synthetic ligands activate PPARα. Synthetic ligands include hypolipidemic drugs such as clofibrate, fenofibrate, ciprofibrate, bezafibrate, nafenopin, and Wy14,643 (pirinixic acid) and industrial phthalate-monoester plasticizers, such as di-(2-ethylhexyl)-phthalate (DEHP), and di-(2-ethylhexyl) adipate (DEHA) and certain pesticides, herbicides, industrial solvents and food flavouring agents (Pyper et al., 2010). Eicosanoid derivatives from the lipoxygenase pathway such as leukotriene B4 (LTB4), oxidized phospholipids and fatty acids are natural ligands for PPARα (Schoonjans et al., 1997; Hostetler et al., 2005). PPARα regulates many target genes and plays a dominant role in fatty acid catabolism and ketone body synthesis in the liver. During starvation, the liver expression of PPARα is increased resulting in enhanced catabolism of fatty acids (Kersten et al., 1999). The fibrate agonists are therapeutically used to lower plasma triglycerides and VLDL by promoting lipid uptake and catabolism and increasing HDL-C (Bays and Stein, 2003; Lefebvre et al., 2006). Other potential therapeutic roles of PPARα agonists include the regulation of atherosclerosis, cardiomyopathies and angiogenesis as well as cancer cell growth (Panigrahy et al., 2008; Crisafulli et al., 2009; Pyper et al., 2010).

PPARγ has many functional roles in different organs and tissues including adipocyte differentiation, glucose metabolism and lipid homeostasis, and is also associated with insulin resistance and atherosclerosis (Spiegelman, 1998). To date, four PPARγ isoforms have been identified in man, due to alternative splicing and differential promoter usage, namely, PPARγ1, γ2 (Fajas et al., 1997), γ3 (Fajas et al., 1998) and γ4 (Sundvold and Lien, 2001). PPARγ1, γ3 and γ4 mRNAs translate into identical proteins, whereas PPARγ2 protein contains an additional 28 N-terminal amino acids (Zhu et al., 1995). PPARγ1 is found in the gut, brain, vascular and immune cells while PPARγ2 is predominantly expressed in adipose tissues (Auwerx, 1999; Desvergne and Wahli, 1999; Rosen and Spiegelman, 2001; Semple et al., 2006) and is extensively studied for its association with diabetes and insulin resistance (Deeb et al., 1998; Douglas et al., 2001; Mori et al, 2001). Prostaglandins like 15-Deoxy-$\Delta^{12,14}$prostaglandin J_2 (15d-PGJ_2), a product of arachidonic acid metabolism and hexadecyl azeloayl phosphatidylcholine (azPC), a product of oxidized low-density lipoproteins are endogenous ligands for PPARγ (Zhang & Young, 2002) while synthetic ligands include the antidiabetic drugs, thiazolidinediones (TZDs), and non-steroidal anti inflammatory drugs (NSAIDs) (Forman et al., 1995; Kliewer et al., 1997; Larsen et al., 2003; Staels and Fruchart, 2005). Examples of TZDs are troglitazone, rosiglitazone, pioglitazone and ciglitazone. PPARγ ligands have the ability to inhibit the proliferation of adipocytes, macrophages and several cancer cells (Debril et al., 2001; Kopelovich et al., 2002). Ligand activation of PAPRγ induces adipogenic differentiation of fibroblasts and preadipocytes (Tontonoz et al., 1994), consistent with its critical role in adipocyte differentiation as well as lipid and glucose homeostasis (reviewed in Schoonjans et al., 1997). PPARγ is also associated with other cellular functions, including macrophage differentiation (Chinetti et al., 1998) and its agonists are considered to have therapeutic potential for cancer treatment since they have been shown to regulate differentiation and growth of various cancer cells (Mueller et al., 1998; Kitamura et al., 1999; Roberts-Thomson, 2000).

PPARδ is ubiquitously expressed in most tissues at varying levels, being highest in the small intestine, colon and liver where it is found predominantly in the nucleus (Fredenrich & Grimaldi, 2005; Girroir et al., 2008; Bishop-Bailey and Bystrom, 2009). PPARδ has been implicated in fatty acid oxidation in many of these tissues (Fredenrich & Grimaldi, 2005). A very high level of PPARδ is also reported in keratinocytes (Girroir et al., 2008). Fatty acids, triglycerides and prostacyclin are endogenous activators of PPARδ (Berger et al., 1999; Bishop-Bailey & Wray, 2003; Peters et al., 2008) and synthetic agonists such as L-165041, GW501516, GW0742 and the antihypertensive drug, Iloprost, selectively binds to PPARδ (Peters et al., 2008; Peters and Gonzalez, 2009). Ligand activation of PPARδ is also reported to increase fatty acid catabolism in skeletal muscle and modulate insulin sensitivity (Grimaldi, 2007). A variety of natural and synthetic compounds activates PPARδ including dietary fatty acids and prostaglandins which can lead to upregulation or repression of target genes (Peters et al., 2008). Modulation of cellular activities by PPARδ can occur via different mechanisms, such as through binding to DNA response elements in association with corepressors (Peters and Gonzalez, 2009). It also interacts with other transcription factors like NF-κB, inhibiting NF-κB-dependent signaling and resulting in anti-inflammatory responses which may be partly mediated by interactions with signal transducer and activator of transcription 3 (STAT3) and extracellular receptor kinase (ERK) 5 (Peters et al., 2008). Due to its high expression in the small and large intestine as well as its anti-inflammatory properties, PPARδ is thought to have some function in gastrointestinal physiology and disease, and may have important functions in the treatment of inflammatory bowel disease (Peters et al., 2008). However, there have been no reports on PPARδ drugs in clinical use, though a few compounds, including GW501516, are currently in clinical trials for dyslipidaemia (Thevis et al., 2010).

PPAR Activation and Transcriptional Control

A number of coregulatory molecules mediate the gene regulatory activities of nuclear receptors including PPARs. These include the p160 family of nuclear coactivators that can enhance ligand-dependent transcriptional activation, and corepressors such as NCoR1 and NCoR2/silencing mediator of retinoic acid and thyroid hormone receptor (SMRT) which function as ligand-dependent or -independent repressors of gene transcription (Rosenfeld and Glass, 2001; Kota et al., 2005). These coregulatory molecules may differentially affect nuclear receptors, guided in part by posttranslational modifications to the receptor (Battaglia et al., (2010). Increased expression of NCoR1 was found to disrupt PPARα/γ signaling by repressing basal expression of target genes in a cell cycle-specific manner, suppressing target gene transctivation and attenuating transrepression ability of PPARα/γ in prostate cancer cells (Battaglia et al., 2010).

Like other nuclear hormone receptors, binding of a ligand activates PPAR, whereby the resulting conformational change triggers the release of NCoR and the recruitment of coactivator proteins, like Nuclear Co-Activators (NCoAs), Cyclic-AMP Response Element Binding Protein/p300 (CBP/p300) and CCAAT/Enhancer binding proteins (C/EBP) (Rosen *et al.*, 2002) that are essential for the initiation of gene transcription, as well as the formation of a heterodimer complex with another nuclear receptor protein, retinoid X receptor (RXR) (Perissi and Rosenfeld, 2005; Genini et al., 2008). The PPAR:RXR heterodimer would then bind to specific DNA sequence motifs termed peroxisome proliferator response elements (PPRE), located within the promoter region of target genes. Both transcriptional activation and repression are mediated by the ligand-binding domain.

PPREs are composed of two direct AGG(A/T)CA repeats separated by a single nucleotide (DR1) found in inducible genes (Ijpenberg et al., 1997; Gervois et al., 1999). Several nuclear receptors mediate their responses through distinct DR sequences ranging from DR1 to DR5 (Mangelsdorf and Evans, 1995). In addition to spacing, subtle differences in the hexad half-site and the 5' extension of these response elements are also important and can have dramatic effects on the activity of a receptor (Mangelsdorf and Evans, 1995).

Nuclear receptors differ in their dimerization and DNA binding properties. Some like ER, glucocorticoid receptor and the androgen receptor, bind as homodimers to their response elements, while others like PPAR, thyroid hormone receptor, vitamin D receptor and all-*trans*-retinoic acid receptor bind as heterodimers with the common partner, RXR (Mangelsdorf and Evans, 1995).

PPAR Phosphorylation

Besides natural and synthetic activating ligands, the transcriptional activities of PPARs are also regulated by post-translational mechanisms including phosphorylation and ubiquination (Blanquart et al., 2003). As phosphoproteins, the activities of PPARs are modulated by their phosphorylation status.

PPARα phosphorylation occurs exclusively on serine residues and treatment of cells with phosphatase inhibitors decreases ligand-induced gene expression (Burns and Vanden Heuvel, 2007). PPARα phosphorylation in the A/B domain is reported to increase insulin response through ERK-mitogen-activated protein kinase (MAPK) pathway and this is associated with increased basal and ligand-induced transcriptional activity (Shalev et al., 1996; Blanquart et al., 2003). The ERK-MAPK signaling is suggested to affect the intra-molecular communication since although phosphorylation occurs in the A/B domain of PPARα, ligand-dependent transctivation but not ligand-independent AF-1 domain is affected (Burns and Vanden Heuvel, 2007). Protein kinase A (PKA) activation also leads to increased PPARα activity but the C domain (DNA-binding) is most strongly phosphorylated compared to A/B and E (ligand-binding) domains (Lazennec et al., 2000). Protein kinase C (PKC) was also reported to increase the transactivated function of PPARα whereby inhibition of PKC signaling decreases PPARα phosphorylation, impairs ligand activation and increases transcription repression function (Blanquart et al., 2004, Gray et al., 2005). Other growth factors such as tumor necrosis factor α (TNFα) and platelet-derived growth factor (PDGF)/ epidermal growth factor (EGF) can also affect PPARα activity, possibly via kinase cascades (Burns and Vanden Heuvel, 2007). A member of the stress-activated kinase family, p38 MAPK, also phosphorylates the A/B domain of PPARα, resulting in increased interaction with the coactivator, PPARγ-coactivator 1α (PGC-1α) and hence, enhancement of ligand-induced transcriptional activity in cardiac myocytes (Barger et al., 2001).

Phosphorylation of serine residues in PPARγ1 or γ2 by MAPK on the other hand, inhibits their transcriptional activities and mutation of the serine residues increases AF-1 transcriptional activity (Hu et al., 1996, Adams et al., 1997; Camp et al., 1999). In humans, both ligand-dependent and -independent transactivating function of PPARγ1 is affected by this phosphorylation (Adams et al., 1997). *In vitro* adipocyte differentiation and cellular growth is dependent on the phsophorylation status which is regulated by growth factors including EGF and PDGF (Camp and Tafuri, 1997). PPARγ phosphorylation may play an important role in obesity and insulin resistance since inhibition of PPARγ phosphorylation was reported to improve insulin sensitivity resulting in increased glucose uptake by muscles (Shao et al., 1998; Rangwala et al., 2003)

Cyclic AMP signaling induces PPARδ-mediated transactivation, resulting in increased basal and ligand-activated activities, influencing heterodimerization and interaction with corepressors (Lazennec et al., 2000; Hansen et al., 2001). As with PPARα, activation of PKA also increases the transcriptional activity of PPARγ and PPARδ isoforms (Lazennec et al., 2000; Blanquart et al., 2003).

PPAR Degradation

The ubiquitin-proteasome degradation system regulates many short-lived proteins via covalent binding of activated ubiquitin to target proteins, followed by degradation by the proteasome (Hodges et al., 1998) hence, rapidly modulating protein levels and key cellular functions such as gene transcription, cell proliferation, cell death and cell cycle progression (Genini et al., 2008).

The ubiquitin-proteasome system also influences the transcriptional activities of several nuclear receptors including PPARs (Floyd and Stephens, 2002;

Blanquart et al., 2003). Downregulation of PPARγ has been reported following ligand-induced transcriptional activation which correlated with increased ubiquitination and proteasome degradation of the receptor (Hauser et al., 2000). Ligand-dependent conformational change in the AF2 domain and/or docking of coactivator and corepressor proteins has been suggested to be important for this response since mutation in this domain failed to induce degradation. In addition, binding of a corepressor protein via the AF2 domain blocked PPARγ degradation (Hauser et al., 2000). Floyd and Stephens (2002) demonstrated in adipocytes that PPARγ1 and PPARγ2 are targeted by the ubiquitin-proteasome pathway under basal conditions, and exposure to interferon γ (IFNγ) results in rapid degradation of these receptors. Inhibition of the proteasome pathway on the other hand, reduces the effect of IFNγ on PPARγ.

PPARα can also be degraded via the ubiquitin-proteasome pathway which helps regulate the transcriptional activities of this receptor. This was confirmed by specific inhibition of PPARα degradation using the highly specific proteasome inhibitor, MG132, resulting in the increase of PPARα transcriptional activation and expression of target genes (Blanquart et al., 2002). However, unlike for PPARγ, ligands for PPARα prevent the degradation of this isoform by decreasing its ubiquitination (Blanquart et al., 2003). Similarly, PPARδ agonists prevent ubiquitination of this receptor in order to maintain its transcriptional activity (Genini and Catapano, 2007).

Phosphorylation may not only affect PPAR transcriptional activity but could also influence PPAR degradation. Inhibitors of MEK kinase prevent PPARγ degradation in adipocytes indicating that PPARγ degradation may be influenced by extracellular signals which activate the intracellular phosphorylation pathways (Floyd and Stephens, 2002). Furthermore, Floyd and Stephens (2002) demonstrated that both ERK1 and 2 play a role in this degradation under basal and IFNγ-mediated conditions. PPARα is also a substrate for glycogen synthase kinase 3 (GSK3) which phosphorylates and inactivates glycogen synthase during glycogen synthesis, and this predominantly occurs in the A/B domain. GSK3 overexpression destabilizes PPARα and facilitates rapid degradation of PPARα by the ubiquitin-proteasome system (Burns and Vanden Heuvel, 2007).

The above studies indicate that the ubiquitin-proteasome degradation system is an important regulator for the cellular levels of PPARα and γ proteins and hence, their function. This degradation pathway may be influenced by the phosphorylation status of these transcription factors. Whether phosphorylation would similarly affect the degradation process of PPARδ is still uncertain.

Reciprocally, PPARs can also promote stabilization or proteasomal degradation of proteins. For example, PPARγ-induced degradation of β-catenin partly mediates the suppression of Wnt signaling required for differentiation of preadipocytes into adipocytes. This action of PPARγ occurs in a GSK3-dependent or –independent manner but requires an active APC-containing destruction complex. Mutations in β-catenin can therefore inhibit PPARγ-induced expression of adipogenic genes (reviewed in Genini et al., 2008). Cyclin D1, a critical component in the regulation of cell cycle, and ERα have also been found to be degraded in breast cancer cells by 15d-PGJ_2 and this downregulation can be blocked by proteasome inhibition using MG132 (Qin et al., 2003). On the other hand, proteosomal degradation of a cyclin-dependent kinase inhibitor, p27, is inhibited by troglitazone leading to inhibition of proliferation of hepatocellular carcinoma cells (Motomura et al., 2004). Activation of NF-κB which is important in inducing inflammatory gene transcription occurs following ubiquitination and degradation of its inhibitory molecule, IκB, leading to translocation of NF-κB into the nucleus (Karin and Ben-Neriah, 2000). The PPARγ ligand, 15d-PGJ2, represses NF-κB activation by inhibiting the proteosomal degradation of IκB (Delerive et al., 2001).

PPAR and Diabetes Mellitus

TZDs were the first group of pharmacological compounds used to address the problem of insulin resistance both in patients with type 2 diabetes mellitus (T2DM) and those with non-diabetic insulin resistance like obesity, polycystic ovary syndrome, aging and metabolic syndrome (Fujiwara and Horikoshi, 2000; Miller, 2003). Several studies have shown that PPARγ activation by TZDs can significantly influence insulin signaling by interfering at various steps of the phosphorylation pathways, ultimately improving insulin sensitivity, enhancing glucose peripheral utilization and lipid metabolism (Leonardini et al, 2009). Treatment of T2DM patients with these drugs lowered both fasting and postprandial glucose levels as well as reducing plasma insulin concentrations (Iwamoto et al, 1991; Suter et al, 1992; Durbin, 2004). In the treatment of metabolic syndrome (otherwise known as syndrome X or insulin-resistance syndrome), TZDs were not only efficient in improving insulin sensitivity but also in reducing circulating triglyceride levels, increasing HDL-C levels, reducing blood pressure and decreasing circulatory plasminogen activator inhibitor 1 (PAI-1) levels (reviewed in Quinn et al., 2008, Hamblin et al., 2009).

Ciglitazone, troglitazone, englitazone, pioglitazone, rosiglitazone and recently netoglitazone (a drug with a dual PPARα/γ agonism) have been tried clinically. Pioglitazone (ACTOS) and rosiglitazone (AVANDIA) have made the greatest contribution to the treatment of T2DM. However, rosiglitazone was recently withdrawn from the European market due to its cardiotoxicity, and troglitazone has previously been withdrawn from the market for its hepatotoxicity (Saha et al., 2010).

PPARα has also been found to play an essential role in glucose homeostasis. Activation of this nuclear receptor upregulates glycerol-3-phosphate dehydrogenase, glycerol kinase and glycerol transport proteins (initiating glucose synthesis during fasting), and promotes the pancreatic glucose-stimulated insulin secretion (Patsouris et al, 2004; Lefebvre et al, 2006). Many clinical trials have shown an improvement in insulin sensitivity following treatment with PPARα agonists like bezafibrate. Fenofibrate has been reported to have renal protective effect with reduced production of (transforming growth factor (TGF) β cytokine that induces renal fibrosis and glomerulosclerosis (Flanders and Burmester, 2003; Chen et al., 2006). In PPARα-knockout mice, the levels of circulating free fatty acids and triglycerides were increased, and extracellular matrix formation and renal inflammation were induced, resulting in more severe/aggravated diabetic nephropathy. Fenofibrate treatment conferred renal protection via reduced albuminuria and glomerular lesions (Park et al., 2006a, 2006b). Based on the 'Fenofibrate Intervention and Event Lowering in Diabetes' (FIELD) study, Sacks (2008) reported that fenofibrate treatment of diabetic patients reduces the rate of progression of albuminuria.

The role of PPARδ in glucose metabolism has not been precisely understood. However, it appears that PPARδ ligands may improve insulin sensitivity by facilitating fatty acid oxidation mainly in the adipose tissue and the skeletal muscles. PPARδ activation inhibits pyruvate dehydrogenase complex through targeting the Pdk4 gene besides the activation of pentose phosphate shunt (Barish et al, 2006; Lee et al, 2006; Jay and Ren, 2007).

Dual PPARα and γ agonists have been developed to increase insulin sensitivity for the prevention of diabetic cardiovascular complications (Etgen et al., 2002) but they are found to have carcinogenic effect and produce cardiovascular risks (Balakumar et al., 2007). PPARα/γ, PPARγ/δ, PPARα/δ dual agonists and PPARα/γ/δ pan agonists are being developed for potential use in diabetic patients (Oyama et al., 2009).

PPAR AND INFLAMMATION

Evidence indicates that PPARα activation reduces inflammation, influencing both acute and chronic inflammatory disorders (Pyper et al., 2010). The anti-inflammatory effects of PPARα are thought to occur via modulation of various inflammatory processes including atherogenesis. Various factors contribute to the development of atherosclerosis including endothelial injury, proliferation of vascular smooth muscle cells and migration of monocytes/macrophages (Takano and Komuro, 2009; Hasegawa et al., 2010)

PPARα activation affects both the initiation and progression of atherosclerosis, resulting in endothelial dysfunction such as reduced leukocyte adhesion and inhibition of transendothelial leukocyte migration as well as inhibition of macrophage foam cell formation (Zandbergen and Plutzky, 2007). This latter effect could occur via modulation of expression of genes of reverse-cholesterol transport (Chinetti et al., 2001; Bighetti et al., 2009; Rotllan et al., 2011), the formation of reactive oxygen species (ROS) (Teissier et al., 2004; Billiet et al., 2008) as well as modification of lipoprotein expression and oxidation (Gbaguidi et al., 2002; Teissier et al., 2004). Accumulation and uptake of oxidized low-density lipoproteins (ox-LDL) by macrophages induces foam cell formation and stimulates the production of proinflammatory molecules by endothelial cells. PPARα also affects local cell immune responses by modulating the expression of genes that control inflammatory responses in endothelial cells, smooth muscle cells and macrophages, which play roles in the pathology of atherosclerosis (Marx et al., 1999; Lee et al., 2000). However, activation of PPARα by ox-LDL resulted in increased production of chemotactic factors for monocytes, MCP-1 and IL-8, conferring its pro-inflammatory effect (Lee et al., 2000). On the other hand, LTB4 serves as an endogenous ligand for PPARα and promotes transcription of genes of the β- and ω-oxidation pathways that regulate the inflammatory response by neutralizing and degrading the LTB4 itself (Devchand et al., 1996). In addition, LTB4- or arachidonic acid-induced inflammatory response is prolonged in PPARα null mice.

Matrix metalloproteinase-12 (MMP-12) is highly expressed in atherosclerotic lesions and ligand activation of PPARα was found to inhibit MMP-12 expression in macrophages treated with the proinflamamtory cytokine, IL-1β (Souissi et al., 2008). Recently, PPARα ligands, Wy14643 and fenofibrate, were shown to suppress the production of PDGF-BB which plays a crucial role in artherosclerosis, as well as the production of inflammatory cytokines induced by PDGF-BB, and may hence exert a protective effect against athersoclerosis

(Hashizume et al., 2010). However, evidence indicates that PPARα deficiency in APoE null mice produces less athersoclerosis instead (Tordjman et al., 2001).

PPARγ activation may also have an important role in the modulation of inflammatory responses. Macrophage stimulation by ox-LDL promotes an oxidative burst upon first contact and elicits ROS formation (Nguyen-Khoa et al., 1999). However, ox-LDL is also a known PPARγ agonist, hence, desensitization of macrophages occurs via PPARγ activation leading to attenuation of ROS formation (Fischer et al., 2002). In line with this, high expression of PPARγ has been observed in foam cells of atherosclerotic lesions (Nagy et al., 1998; Tontonoz et al., 1998).

Observation of PPARγ expression in vascular and inflammatory cells (Blaschke et al., 2006) may also imply that PPARγ could regulate atherosclerosis-associated genes such as those of vascular cell adhesion molecule-1 (VCAM-1) and inflammatory cytokines. Concordantly, PPARγ ligands were reported to inhibit the expression of VCAM-1 and intracellular adhesion molecule-1 (ICAM-1) as well as the production of chemokines (Pasceri et al., 2000). The expression of macrophage-induced matrix metalloproteinase, MMP-9, was also decreased leading to the inhibition of vascular smooth muscle cell migration (Marx et al., 1998; Ricote et al., 1998). VCAM-1 expresssion is increased by NFκB activation. VCAM-1 and IκBα, which indirectly regulates VCAM-1 expression via NFκB, regulate inflammatory responses in atherosclerosis. IκBα is induced by PPARα activation (Delerive et al., 2000) and interestingly, the PPARγ agonist, pioglitazone, repressed inflammation via reduction in VCAM-1 expression and increased hepatic IκBα expression in a PPARα-dependent manner both *in vitro* and *in vivo* in mice (Orasanu et al., 2008)

PPAR AND THE IMMUNE SYSTEM

All PPAR subtypes have been shown to be expressed in immune cells including macrophages, B and T lymphocytes, and dendritic cells. PPARs, particularly PPARα and PPARγ, have been shown to regulate inflammation and the immune response (Moraes et al., 2006). PPAR ligands have been found to attenuate the activity of many autoimmune disease processes like inflammatory bowel disease in animal models (Guri et al, 2010), arthritis (Afif et al, 2007), psoriasis (Romanowska et al, 2008) and experimental autoimmune encephalomyelitis (EAE) (Defaux et al., 2009) as well as to reduce inflammation in patients with mild and moderate ulcerative colitis (Savoye, 2008). Our own observations on autoimmune diabetes suggest that PPARγ may be important in suppressing the autoimmune process, and that its role may be influenced by the sex of the mouse model (non-obese diabetic strain) used (Yaacob et al., 2009; 2010).

PPARα activation through ligand binding has been found to inhibit the production and release of several proinflammatory molecules by interfering with the AP1 and NF-κB signaling pathway (Jiang et al., 1998; Ricote et al., 1998; Shu et al, 2000; Neve et al, 2001). PPARα agonists like gemfibrozil and fenofibrate, which are well-established anti-dyslipidemic drugs, have been shown to inhibit IL-2, TNFα, and IFNγ production by activated $CD4^+$ T cells and to induce IL-4 production in splenocytes. In addition, these agonists may also inhibit inflammatory immune response by regulating the action of NF-κB by promoting the expression of IκBα (Gocke et al, 2009). Inactivation of NF-κB may also repress the expression of inflammatory mediators induced by extracellular inflammatory stimuli like VCAM-1 and the TNFα-induced ICAM-1 (Blanquart et al., 2003). PPARα knockout mice have been shown to exhibit a prolonged inflammatory response when treated with LTB4, the endogenous ligand for PPARα (Narala et al., 2010)

The most well studied PPAR in the suppression and regulation of immune and inflammatory responses is PPARγ (Zhang & Young, 2002; Bishop-Bailey & Wray, 2003). Activation of PPARγ inhibits the production of inflammatory cytokines, the proliferation of mitogen-activated T cells (Clark et al., 2000; Yang et al., 2000; Wang et al., 2001; Cunard et al., 2002), the activation of macrophages (Jiang et al., 1998; Rotondo and Davidson, 2002) and the activation of of inflammatory transcription factors like NF-κB, AP-1 and STAT1, which activates the expression of inflammatory proteins like iNOS, gelatinase B and SR-A (Ricote et al., 1998; Wang et al., 2001). Besides inhibiting inflammatory responses, it also promotes the expression of anti-inflammatory cytokines like IL-4, IL-5 and IL-13 (Zhang & Young, 2002).

In animal studies, the onset of inflammatory bowel disease in IL-10 deficient mice can be delayed by treatment of rosiglitazone which reduced the level of Th1 cytokines as well as iNOS (Lytle *et al.*, 2005). The autoimmune systemic lupus erythematosus MRL/*lpr* mouse kidney and mesangial cells (isolated from kidney glomerulus) were observed to have a decreased expression of PPARγ compared to BALB/c control mice, suggesting a role for PPARγ in modulating the lupus inflammatory response of the kidney (Crosby *et al.*, 2006).

There are several suggested mechanisms of PPARγ action in immunoregulation. Yang et al. (2000) demonstrated the binding of PPARγ with

NFAT, a transcription factor regulating the IL-2 gene promoter. The binding prevents the expression of IL-2, which is required for T cell proliferation. PPARγ inhibits expression of the T-bet gene, which is essential for the differentiation of CD4 T cells into the Th1 subtype and in contrast promotes the expression of GATA3, a transcription factor analogous to T-bet, which differentiates CD4 T cells into the Th2 subtype (reviewed by Zhang & Young [2002]). Suppression of the inflammatory response by PPARγ activation may also be achieved through induction of apoptosis in immune cells (Jiang et al., 1998, Harris et al., 2002).

PPARγ activation has also been shown to regulate inflammatory responses by interacting with NF-κB, AP-1, C/EBP, and STAT on the promoters of target genes (Zelcer and Tontonoz, 2005). PPARγ enhances the differentiation of monocytes into the more anti-inflammatory M2 macrophage phenotype utilizing Th2 cytokines like IL-4 and IL-13 (Bouhlel et al., 2007; Odegaard, et al, 2007). PPARγ has also been found to play a crucial role in the regulation of the innate immune response in both humans and animal models. PPARγ activation enhances phagocytosis of apoptotic neutrophils and suppression of neutrophil chemotaxis (Asada et al, 2004; Majai et al, 2007).

PPARγ ligands act as a repressor of the major histocompatibility complex class II (MHC-II)–mediated T-cell activation by inhibition of IFNγ–induced MHC-II expression (Kwak et al, 2002). PPARγ activation has also been shown to delay the onset of eczematous skin lesions with concomitant reduction in mast cells, CD4 and CD8 lymphocyte infiltration. Moreover, allergen-specific IgE and IgG1 responses were reduced. It has been suggested that PPARγ ligand treatment inhibits not only systemic allergic immune response, but also local allergen-mediated dermatitis (Demerjian et al, 2006; Dahten et al, 2008). Clinical trials have shown that the administration of PPARγ and PPARα ligands (glitazones and fibrates) significantly reduced the risk of atherosclerosis, a process that is known to involve chronic immune-inflammatory reactions.

PPARδ has been suggested to suppress inflammation (Bishop-Bailey & Bystrom, 2009) as observed in experimental lupus-like autoimmune kidney damage (Mukundan et al, 2009) as well as in EAE (Polak et al., 2005). PPARδ has recently been suggested to be an essential regulator of autoimmune inflammation via inhibition of IFNγ and IL-12 (Dunn et al., 2010). Currently there is a lot of interest on PPARδ and its role in the immune response, and reports in this area is eagerly awaited.

PPAR AND CANCER

Drugs that activate PPARs are now widely studied for their ability to inhibit tumor cell growth. PPARγ, the best characterized of the PPARs, is overexpressed in various tumors including breast (Elstner et al., 1998), colon (Chen et al., 2002) and lung (Theocaris et al., 2002). As such, the involvement of PPARγ in carcinogenesis is intensely investigated although the exact role of this transcription factor has yet to be established. Many studies support the role of PPARγ as a tumor suppressor, whilst others have shown that it can promote tumor formation (Koeffler, 2003).

PPARγ ligands have been shown to inhibit growth of various cancer cells (Elstner et al., 1998; Mueller et al., 1998; Sarraf et al., 1998; Debril et al., 2001; Goke et al., 2001; Chang and Szabo, 2002; Harris and Phipps, 2002; Kopelovich et al., 2002; Chen et al., 2003; 2005; Li et al., 2005). 15d-PGJ_2, is reported to show anti-inflammatory, anti-proliferative and pro-apoptotic properties (Strauss and Glass, 2001). Activation of PPARγ following treatment with 15d-PGJ_2 has been reported to inhibit tumor cell growth both *in vitro* and *in vivo* in a variety of tissues, including breast, prostate, colon, lung, bladder and esophagus (Harris et al., 2002). 15d-PGJ_2 has been shown to be a potent inducer of caspase-mediated apoptosis in a variety of cells (Bailey and Hla, 1999, Sato et al., 2000; Rohn et al., 2001). However, 15d-PGJ_2 only transiently increased PPARγ1 and γ2 expression in breast cancer cells and pharmacologic inhibition of PPARγ indicates that this apoptotic activity of 15d-PGJ_2 occurs independently of the receptor (unpublished observations), consistent with previous reports on PPARγ-independent apoptosis in prostate and bladder cancer cells (Chaffer et al., 2006).

Suh et al., (1999) first reported on the prevention of experimental breast cancer by a PPAR ligand, GW7845, and activation of PPARγ by troglitazone was later shown to be protective against DMBA-induced breast tumor development and this chemopreventive activity was potentiated by the presence of a ligand for RXR, LG10068 (Mehta et al., 2000). Treatment of MCF-7 breast cancer cells with rosiglitazone activates PPARγ which in turn activates p53 gene and causes apoptosis through a cascade of downstream signaling which includes expression of p21 and activation of caspase-9 (Bonofiglio et al., 2006).

Various studies reported on signaling cross-talk between PPAR and ER (Keller et al., 1995; Wang and Kilgore, 2002; Bonofiglio et al., 2005; Suzuki et al., 2006; Kim et al., 2007) whereby the PPAR/RXR heterodimer inhibited transactivation of ER through competition for ER response element (ERE) binding (Keller et al., 1995). The signal cross talk is suggested to exist bidirectionally between PPARγ and ER in breast cancer cell lines whereby the expression of either ERα or ERβ reduces both basal and stimulated PPARγ-mediated reporter activity (Wang and Kilgore, 2002). A strong association of PPARγ and ERα status in breast cancer patients has been found (Suzuki et al., 2006). Our own study showed that 17β-estradiol enhanced the effect of 15d-PGJ2 mediated killing of ER-positive but not ER-negative cell lines, providing further evidence of the association between PPARγ and ERα in breast cancer (unpublished observations).

Inhibition of colorectal adenocarcinoma cell growth has also been reported. The well differentiated HT-29 and poorly differentiated COLO-205 cell growth was potently inhibited by ciglitazone via caspase-dependent apoptosis in a time-dependent manner, although this may be independent of PPARγ. It was also observed that the expression of PPARγ1 but not PPARγ2 genes was significantly downregulated with a concomitant decrease in PPARγ protein levels (Yaacob et al., 2008). Treatment of HT-29 with 15d-PGJ_2 and rosiglitazone also results in apoptosis of the tumor cells. Treatment with rosiglitazone could increase the expression of PTEN, a tumor suppressor gene and the increase is reversible when the cells were treated with GW9662, a PPARγ selective inhibitor (Chen et al., 2005).

Many factors seem to contribute to the expression and regulation of PPARγ activity, making the identification of specific role(s) of PPAR in regulating cancer cell growth difficult. The stage of development of specific tumor cells has also been suggested to influence the physiological consequence of PPARγ activation (Allred and Kilgore, 2005). These authors also demonstrated that rosiglitazone, ciglitazone and 15d-PGJ_2 differentially modulate PPRE reporter activity in various breast, colon and lung cancer cell lines suggesting that cancer cells respond differentially to individual PPARγ ligands depending on the cell characteristics.

The hepatocarcinogenic effect of peroxisome proliferators has long been recognized in rodents (Rao and Reddy, 1996). However, the lack of direct genotoxicity of peroxisome proliferators has led to the suggestion that the hepatocarcinogenicity is linked to metabolic disturbances resulting from the suppression or delay in apoptosis (Bursch et al., 1992; Bayly et al.,1994), oxidative stress (Reddy and Lalwani,1983; Conway et al., 1989; Kasai et al., 1989) or tumor promotional activity (Marsman et al., 1988; Cattley and Popp, 1989; Kraupp-Grasl et al., 1990; Tanaka et al., 1992), through modulation of cell growth signals (Reddy and Chu, 1996). As such, oxidative stress together with hepatocellular proliferation may be responsible for the development of hepatocellular carcinomas in rodents following chronic exposure to peroxisome proliferators, resulting from sustained receptor activation (Peters et al., 2005). This is supported by a concordant induction of the expression of genes encoding proteins involved in DNA repair pathways which is crucial for the repair of oxidized DNA lesion (Rusyn et al., 2004; Woods et al., 2007). Furthermore, PPARα knockout mice were found to have impaired tumor growth (Kaipainen et al., 2007). It was also recently found that PPARα activation induced DNA damage repair network. However, the exact mechanism of induction is unclear but the involvement of oxidative stress mechanism was suggested (Qu et al., 2010). These observations could explain the possible mechanism leading to peroxisome proliferator-induced hepatocarcinogenicity.

On the other hand, a tumor suppressive role of PPARα is suggested since it can interfere with the metabolic pathways in cancer cells. It has previously been reported that carcinogenesis is often linked to mitochondrial dysfunction (Alirol and Martinoue, 2006). PPARα metabolically targets the energy balance of cancer cells by inhibiting glycolysis, fatty acid and lipid syntheses and promoting fatty acid β-oxidation (Grabacka and Reiss, 2008). PPARα also exerts an anti-inflammatory response and upregulates uncoupling proteins (UCP) which function in mitochondrial thermogenesis and reduce ATP biosynthesis. UCP2 has been reported to control proliferation by promoting fatty acid oxidation and limiting glycolysis-derived pyruvate catabolism (Pecqueur et al., 2008), hence suggesting an anticancer effect of PPARα via upregulation of UCP. At the same time, respiratory ROS formation is attenuated which is supported by the observation that mice deficient in PPARα have a higher level of oxidative damage in cardiac muscle (Guellich et al., 2007). Under conditions of limited nutrient availability such as in the tumor microenvironment, PPARα together with AMPK (a potent inhibitor of Akt-induced glycolysis), can repress oncogenic Akt activity, inhibit cellular proliferation and drive cancer cells into metabolic catastrophe (Grabacka and Reiss, 2008). Additionally, PPARα agonists have been shown to inhibit proliferation of cancer cells *in vitro* and demonstrated *in vivo* antitumor activity in murine models (Saidi et al., 2006; Yokoyama et al., 2007; Panigrahy et al., 2008; Chang et al., 2010). The tumor suppressive effect of fenofibrate in carcinogen-induced oral cancer mouse model was suggested to be attributed to the crosstalk between PPARα activation and epidermal growth factor receptor (EGFR) expression and/or cyclooxygenase 2 (COX-2) regulatory pathway (Chang et al., 2010).

Muzio et al (2006) suggested that the apoptotic effect of peroxisome proliferators involves PPARα activation. It was further shown that PPARα activation directly regulates *c-myc* gene expression and induces apoptosis by increasing the expression of pro-apoptotic BAD, while PPARγ activation is important for inhibition of cell proliferation without induction of cell death (Maggiora et al., 2010).

Interests in the role of PPARδ in cancer are relatively new compared to those of PPARγ and PPARα. The role of PPARδ in tumorigenesis is controversial. While PPARδ has been suggested to promote apoptosis in lung cancer (Fukumoto et al, 2005), its role in colon cancer development is controversial.

PPARδ is highly expressed in colorectal cancers (He et al., 1999; Shureiqi et al., 2003; Zuo et al., 2006) and genetic disruption of PPARδ in colon cancer cell lines has resulted in decreased growth of xenografts (Park et al., 2001). Studies suggested that PPARδ ligands could stimulate proliferation of long-term cultured endothelial cells (Stephen et al, 2004) or chemically-induced colonic cancer in mice (Zuo et al., 2009). Stephen et al. (2004) examined the effects of potent and highly selective PPARδ agonists on a range of human epithelial cell lines and found that PPARδ activation can increase the growth of breast and prostate cancer cells under conditions of hormonal deprivation. A mouse model with targeted deletion of PPARδ in its colonic epithelium, had drastically reduced incidence of azoxymethane-induced colon tumors compared to wild-type mice suggesting a tumor-promoting role of this receptor (Zuo et al., 2009). These studies lend support to the notion that PPARδ may promote cancer development.

Several studies indicate that NSAIDs can suppress colorectal carcinogenesis (Lanas, 2009; Cooper et al., 2010; Din et al., 2010) and inhibition of human colon cancer cell growth by NSAIDS has been attributed to downregulation of PPARδ expression (Shureiqi et al., 2003; Liou et al., 2007; Peters et al., 2008; Wu and Liou, 2009) and inhibition of COX 2-derived prostacyclin (PGI_2, a ligand for PPARδ) (Wu and Liou, 2009). The NSAIDs, sulindac and indomethacin, were shown to downregulate PPARδ activity via disruption of the DNA binding ability of PPARδ/RXR heterodimers (He et al., 1999).

On the other hand, growing evidence indicates that PPARδ activation promotes terminal differentiation and inhibits growth of a variety of cancer cells (Schmuth et al., 2004; Planavila et al., 2005; Aung et al., 2006; Kim et al., 2006; Marin et al., 2006; Burdick et al., 2007; Teunissen et al., 2007). Disruption of PPARδ resulted in increased colon tumorigenesis in azoxymethane-induced mice model compared to normal PPARδ expressing mice (Harman et al., 2004). These PPARδ knockout mice were also more susceptible in developing colonic polyps when treated with azoxymethane compared to normal mice. Increased PPARδ expression in colon cancer cell lines was suggested to contribute to indomethacin-induced inhibition of proliferation of these cells (Foreman et al., 2009). In line with this, silencing of PPARδ using RNA interference strategy resulted in significant promotion of proliferation of the colorectal cancer cells, HCT116, as a result of cell cycle dysregulation (Yang et al., 2008).

Taken together, the jury's still out on the potential use of PPARδ ligands as anti-cancer agents. Detailed understanding of various interacting pathways including those involving other PPAR subtypes need to be clarified before selected PPAR ligands could be effectively used for this purpose.

PPAR POLYMORPHISM

The most common PPARγ polymorphism is the substitution of proline to alanine (Pro12Ala) in PPARγ2, first identified by Yen et al. (1997). This occurs as a result of missense mutation in exon B of the adipocyte-specific γ2 isoform with varying frequencies in different ethnic populations (Yen et al., 1997). The highest frequency was found in Caucasians (12%), followed by Mexican Americans (10%), West Samoans (8%), African Americans (3%) and Chinese (1%). A much higher prevalence of this polymorphism (28%) was later reported among Danish Caucasians (Frederikson et al., 2001).

Pro12 variant is reported to be a risk allele for diabetes (Altshuler et al., 2000). Pro12Ala polymorphism on the other hand, is shown to enhance weight loss (Vogels et al., 2005; Goyenechea et al., 2006) and is associated with resistance to development of T2DM (Altshuler et al., 2000; Mori et al., 2001) with a significant increase in circulatory insulin clearance as a result of reduced free fatty acid delivery during hyperinsulinaemia (Tschritter et al., 2003), greater insulin sensitivity and reduced risk for T2DM (Hara et al., 2000; Tonjes & Stumvoll, 2007). The Ala12 allele has also been reported to reduce the risk of myocardial infarction (Ridker et al, 2003).

The significance of association between Pro12Ala polymorphism and T2DM, insulin resistance and obesity has however been controversial. While some studies reported on the absence of association between this polymorphism and metabolic syndrome (Mancinni et al., 1999; Rhee et al., 2007; Badii et al., 2008) others have found that the mutation is positively associated with decreased risk of T2DM in various populations, with greater insulin sensitivity (Gonzalez Sanchez et al., 2002; Rosmond et al., 2003; Ghoussaini et al., 2005; Tavares et al., 2005; Meshkani et al., 2007).

A 10-year (1998 to 2008) meta- analysis covering investigations on Caucasian, European, Asian, North American and South American populations revealed great variability in the frequency of polymorphism (0.04 to 0.55) with the lower frequencies being associated with lower average body mass index (BMI) (Huguenin and Rosa, 2010). It was suggested that the Ala allele has a protective role in lowering the development of T2DM in Caucasians (but not in some other populations) and greater insulin sensitivity in overweight subjects

The silent C to T mutation in exon 6 is the second most common polymorphism of PPARγ gene. It was identified by Meirhaegue and colleagues in 1998 and although it does not lead to amino acid change, this polymorphism has been associated with bone mineral density, obesity and glucose tolerance (Ogawa et al., 1999; Valve et al., 1999; Poulsen et al., 2003) as well as the pathogenesis of atherosclerosis (Wang et al., 1999). Gene-to-gene interactions have been suggested between Pro12Ala and C161T polymorphisms with regard to obesity (Dongxia et al., 2008). Although association of these polymorphisms with metabolic syndrome was not observed they were significantly associated with insulin resistance. The authors suggested linkage disequilibrium with other polymorphism sites and certain promoter regions or introns affecting PPARγ expression as possible mechanism for the improved insulin sensitivity rather than direct functional modification. Excessive PPARγ activation causes body triglyceride redistribution and increased glucose uptake by muscles, hence improving insulin sensitivity. On the other hand, PPARγ downregulation could decrease leptin expression and lipid production, while increasing fatty acid burning and hence decreasing liver and muscular triglyceride content (Dongxia et al., 2008).

Conflicting results have been obtained regarding the association of PPARγ polymorphism with cancer. Most of the studies have been conducted in colorectal cancer. The study of Murtaugh et al (2005) showed that Ala12 allele is associated with reduced risk of colon cancer but increased risk of rectal cancer. On the other hand, Theodoropoulos et al. (2006) reported increased colon cancer risk. A recent meta-analysis of nine independent studies showed weak association with colon cancer with no evidence of association with colorectal or rectal cancer (Lu et al., 2010). Similarly, Pro12Ala variant is also associated with increased risk of gastric cancer in the Chinese population but this association was not observed in *H. pylori*-negative subjects (Liao et al., 2006).

Other less frequent PPARγ genetic variants include the substitution of proline to glutamine (Pro115Gln) resulting in gain of function of this receptor associated with obesity but not insulin resistance (Ristow et al., 1998), and Val290Met and

Pro467Leu mutations which result in defective transcription factor function observed in a few individuals with severe insulin resistance, T2DM and hypertension (Barroso et al., 1999).

More data is needed on the outcome of PPAR genetic variations in various well characterized populations as this would have important implications on health and disease protection or risk. Importantly the possible influence of the environment and/or lifestyle on gene-gene interactions may determine the health outcome and response to therapy.

CROSS-TALK BETWEEN PPARS

PPARs have distinguishable biological functions conferred by their tissue-specific pattern of expression. But they seem to have overlapping therapeutic effects probably due to their high structural similarity and hence, there is no exclusive or specific ligand for each subtype. This property can be valuable for the development of multi-acting PPAR drugs for a variety of diseases but this can also pose problems due to the differential responses resulting from simultaneous activation of the different subtypes. This is particularly important since different PPAR subtypes exhibit opposing biological roles and thus understanding molecular interactions between PPARs can help in the development of drugs that target multiple functions in the cell.

Functional cross-talks between the subtypes are thought to exist in controlling their expression but the mechanism involved is not yet clear. PPAR subtypes may compete for the same binding site since they do not have DNA binding specificity (Lemay and Hwang, 2006; Ricote and Glass, 2007) but the resulting activity may depend on the presence of endogenous PPAR ligands and the interactions with co-regulators (Michalik and Wahli, 2008). In fact, physical and functional interaction occurs between NCoR and all three PPAR subtypes with the ranking order of PPARδ > PPARγ > PPARα (Krogsdam et al, 2002). As such, corepressor as well coactivator protein binding would strongly determine the transcription and function of PPAR target genes. In fact specificity of known ligands for PPAR subtypes may now be questioned based for example on the observation that the PPARγ ligand, pioglitazone, could also activate the LBD of PPARα and induce expression of PPARα target genes (Orasanu et al., 2008).

The presence of cross-talks between PPARs following ligand activation is also implicated when combined applications of PPAR agonists produce responses that differ from the total effects of individual agonists. This was evident in the control of expression and activity of a key enzyme in prostaglandin synthesis, COX-2, in rat brain astrocytes (Aleshin et al., 2009). PPARγ was reported to have a positive influence on PPARδ expression and activity while PPARα exerted a negative influence on the latter. Thus the activity of each PPAR subtype may be controlled by not only its absolute level but also the ratio of all subtype levels such that the responses produced by combined PPAR agonists differ from the cumulative effect of individual agonists (Aleshin et al., 2009).

As seen with some PPARγ ligands, reports on the association of dual PPARα/γ agonists such as tesaglitazar and muraglitazar with cardiovascular risks and carcinogenicity have been put forward (Nissen et al., 2005; Balakumar, 2007). Understanding the regulation of PPAR and the interactions between PPAR subtypes is therefore important for new drug development.

SUMMARY AND FUTURE OUTLOOK

Since their first discovery, PPARs have been shown to be involved in a multitude of overlapping cellular functions. Interactions between PPAR subtypes influence their transcriptional regulation and their target gene expression; and their resulting actions in different tissues would depend on their tissue-specific expression, interactions with different regulatory proteins and the presence of endogenous ligands. The involvement of PPARs in the pathogenesis of diseases such as diabetes, autoimmune disease and cancer has drawn intense interests in developing pharmaceutical targets against these transcription factors. However, the development of such drugs has been impeded by findings of toxicity in the clinical setting, which resulted in the withdrawal of several drugs previously thought to be safe. This fact compels us to understand the complexities of cellular pathways involved in the pleotropic actions of PPARs. Safe and efficacious pan or dual PPAR agonists of natural origins such as abscisic acid, punicic acid and catalpic acid are currently being analyzed for their potential development as therapeutics for obesity, diabetes and inflammation (Bassaganya-Riera et al., 2011).

ACKNOWLEDGMENTS

We thank Kenny Goh and Amjed Hazim Abed for gathering some of the information for this review. Part of the authors' study described in this review was supported by the respective FRGS and IRPA EA grants provided by the Ministry of Higher Education and the Ministry of Science, Technology and Innovation, Malaysia.

References

Adams M, Reginato MJ, Shao D, Lazar MA, Chatterjee VK. Transcriptional activation by peroxisome proliferator-activated receptor gamma is inhibited by phosphorylation at a consensus mitogen- activated protein kinase site, *J Biol Chem.* 1997; 272 (8): 5128–32

Afif H, Benderdour M, Mfuna-Endam L, Martel-Pelletier J, Pelletier JP, Duval N, Fahmi H. Peroxisome proliferator-activated receptor γ1 expression is diminished in human osteoarthritic cartilage and is downregulated by interleukin-1β in articular chondrocytes. *Arthritis ResTher.* 2007; 9:R31

Aleshin S, Grabeklis S, Hanck T, Sergeeva M, Reiser G. Peroxisome proliferator-activated receptor (PPAR)-gamma positively controls and PPARalpha negatively controls cyclooxygenase-2 expression in rat brain astrocytes through a convergence on PPARbeta/delta via mutual control of PPAR expression levels. *Mol Pharmacol.* 2009; 76(2):414-24.

Alirol E, Martinoue JC. Mitochondria and cancer: is there a morphological connection? *Oncogene.* 2006; 25(34):4706–16

Allred CD, Kilgore MW. Selective activation of PPARγ in breast, colon and lung cancer cell lines. *Mol Cell Endocrinol.* 2005; 235:21-9

Altshuler D, Hirschhorn JN, Klannemark M, Lindgren CM, Vohl MC, Nemesh J, Lane CR, Schaffner SF, Bolk S, Brewer C, Tuomi T, Gaudet D, Hudson TJ, Daly M, Groop L, Lander ES.The common PPARgamma Pro12Ala polymorphism is associated with decreased risk of type 2 diabetes. *Nat Genet.* 2000; 26:76–80

Asada K, Sasaki S, Suda T, Chida K, Nakamura H. Antiinflammatory roles of peroxisome proliferator-activated receptor gamma in human alveolar macrophages. *Am J Respir Crit Care Med.* 2004; 169:195–200.

Aung CS, Faddy HM, Lister EJ, Monteith GR, Roberts-Thomson SJ. Isoform specific changes in PPAR alpha and beta in colon and breast cancer with differentiation. Biochem. *Biophys. Res. Commun.* 2006; 340, 656–60

Auwerx J. PPARgamma, the ultimate thrifty gene. *Diabetologia* 1999; 42:1033-49

Badii R, Bener A, Zirie M, Al-Rikabi A, Simsek M, Al-Hamaq AO, Ghoussaini M, Froguel P, Wareham NJ. Lack of association between the Pro12Ala polymorphism of the PPAR-gamma 2 gene and type 2 diabetes mellitus in the Qatari consanguineous population. *Acta Diabetol.* 2008; 45(1):15-21

Bailey DB, Hla T: Endothelial cell apoptosis induced by peroxisome proliferator-activated receptor ligand 15d-PGJ$_2$. *J Biol Chem.* 1999; 274:17042-8

Balakumar P, Rose M, Ganti SS, Krishan P, Singh M. PPAR dual agonists: are they opening Pandora's Box? *Pharmacol Res.* 2007; 56:91–8

Barger PM, Browning AC, Garner AN, Kelly DP. p38 mitogen-activated protein kinase activates peroxisome proliferator-activated receptor alpha: a potential role in the cardiac metabolic stress response. *J Biol Chem.* 2001; 276:44495–501

Barish G, Narkar V, Evans R. PPARδ/β: a dagger in the heart of metabolic syndrome. *J Clin Invest.* 2006; 116:590-7

Barroso I, Gurnell M, Crowley VE, Agostini M, Schwabe JW, Soos MA, Maslen GL, Williams TD, Lewis H, Schafer AJ, Chatterjee VK, O'Rahilly S. Dominant negative mutations *Nature* 1999; 402(6764):880-3.

Bar-Tana J. (2001). Peroxisome Proliferator-activated receptor-γ activation and its consequences in humans. *Toxicol Lett.* 2001; 120:9-19

Bassaganya-Riera J, Guri AJ, Hontecillas R. Treatment of obesity-related complications with novel classes of naturally occurring PPAR agonists. *J Obes.* 2011; Article ID 897894, 7 pages doi:10.1155/2011/897894

Battaglia S, Maguire O, Thorne JL, Hornung LB, Doig CL, Liu S, Sucheston LE, Bianchi A, Khanim FL, Gommersall LM, Coulter HS, Rakha S, Giddings I, O'Neill LP, Cooper CS, McCabe CJ, Bunce CM, Campbell MJ. Elevated NCOR1 disrupts PPARalpha/gamma signaling in prostate cancer *Carcinogenesis.* 2010; 31(9):1650-60

Bayly AC, Roberts RA, Dive C. Suppression of liver cell apoptosis *in vitro* by the non-genotoxic hepatocarcinogen and peroxisome proliferator nafenopin. *J. Cell Biol.* 1994; 125(1):197-203

Bays H, Stein EA. Pharmacotherapy for dyslipidaemia--current therapies and future agents *Expert Opin Pharmacothe*, 2003; 4:1901-38

Bensinger SJ, Tontonoz P. Integration of metabolism and inflammation by lipid-activated nuclear receptors. *Nature* 2008; 454:470–7

Berger J, Leibowitz MD, Doebber TW, Elbrecht A, Zhang B, Zhou G, Biswas C, Cullinan CA,Hayes NS, Li Y, Tanen M, Ventre J,Wu MS, Berger GD,Mosley R, Marquis R, Santini C, Sahoo SP, Tolman RL, Smith RG, Moller DE. Novel peroxisome proliferator-activated receptor (PPAR) gamma and

PPARdelta ligands produce distinct biological effects. *J Biol Chem*. 1999; 274(10):6718–25

Bighetti EJ, Patrício PR, Casquero AC, Berti JA, Oliveira HC. (2009): Ciprofibrate increases cholesteryl ester *Lipids Health Dis*. 2009; 8:50-9

Billiet L, Furman C, Cuaz-Pérolin C, Paumelle R, Raymondjean M, Simmet T, Rouis M. Thioredoxin-1 and its natural inhibitor *J Mol Biol*. 2008; 384(3):564-76

Bishop-Bailey D, Wray J. Peroxisome proliferator-activated receptors: A critical review on endogenous pathways for ligand generation. *Prostaglandins Other Lipid Mediat*. 2003; 5490:1-22

Bishop-Bailey D, Bystrom J. Emerging roles of peroxisome proliferator-activated receptor-beta/delta in inflammation. *Pharmacol Ther*. 2009; 124(2):141-50

Blanquart C, Barbier O, Fruchart JC, Staels B Glineur C. Peroxisome proliferator-activated receptor alpha (PPARalpha) turnover by the ubiquitin-proteasome system controls the ligand-induced expression level of its target genes. *J Biol Chem*. 2002; 277:37254-9

Blanquart C, Barbier O, Fruchart JC, Staels B, Glineur C. Peroxisome proliferator-activated receptors: regulation of transcriptional activities and roles in inflammation. *J Steroid Biochem Mol Biol*. 2003; 85(2-5):267-73

Blanquart C, Mansouri R, Paumelle R, Fruchart JC, Staels B, Glineur C. The protein kinase C signaling pathway regulates a molecular switch between transactivation and transrepression activity of the peroxisome proliferator-activated receptor alpha. *Mol Endocrinol*. 2004; 18:1906-18

Blaschke F, Caglayan E, Hsueh WA. Peroxisome proliferatoractivated receptor gamma agonists: their role as vasoprotective agents in diabetes. *Endocrinol Metab Clin North Am* 2006;35:561–74.

Bonofiglio D, Gabriele S, Aquila S, Catalano S, Gentile M, Middea FG, Ando S. Estrogen receptor binds to peroxisome proliferator–activated receptor response element and negatively interferes with peroxisome proliferator–activated receptor signaling in breast cancer cells. *Clin Cancer Res*. 2005; 11:6139-47

Bonofiglio D, Aquila S, Catalano S, Gabriele S, Belmonte M, Middea E, Qi H, Morelli C, Gentile M, Maggiolini M, Andò S. Peroxisome proliferator-activated receptor-gamma activates p53 gene promoter binding to the nuclear factor-kappaB sequence in human MCF7 breast cancer cells. *Mol Endocrinol*. 2006; 20(12):3083-92

Bouhlel MA, Derudas B, Rigamonti E, Dievart R, Brozek J, Haulon S, Zawadzki C, Jude B, Torpier G, Marx N, Staels B, Chinetti-Gbaguidi G. PPAR gamma activation primes human monocytes into alternative M2 macrophages with anti-inflammatory properties. *Cell Metab.* 2007; 6:137-43

Braissant O, Foufelle F, Scotto C, Dauca M, Wahli W. Differential expression of peroxisome proliferator-activated receptors (PPARs): tissue distribution of PPAR-alpha, -beta and –gamma in the adult rat. *Endocrinology.* 1996; 137:354–66

Burdick AD, Bility MT, Girroir EE, Billin AN, Willson TM, Gonzalez FJ, Peters JM. Ligand activation of peroxisome proliferator-activated receptor-beta/delta(PPARbeta/delta) inhibits cell growth of human N/TERT-1 keratinocytes. *Cell Signal.* 2007; 19:1163–71

Burns KA, Vanden Heuvel JP. Modulation of PPAR activity via phosphorylation. *Biochim Biophys Acta.* 2007; 1771(8): 952-60

Bursch W, Oberhammer F, Schulte-Hermann R. Cell death by apoptosis and its protective role against disease. *TIPS.* 1992; 13:245-51

Camp HS, Tafuri SR. Regulation of peroxisome proliferatoractivated receptor gamma activity by mitogen-activated protein kinase, *J. Biol. Chem.* 1997; 272: 10811–6

Camp HS, Tafuri SR, Leff T. c-Jun N-terminal kinase phosphorylates peroxisome proliferator-activated receptor-g1 and negatively regulates its transcriptional activity, *Endocrinology.* 1999; 140:392–7

Cattley RC, Popp JA. Differences between the promoting activities of the peroxisome proliferator WY-14,643 and phenobarbital in rat liver. *Cancer Res.* 1989; 49:3246-51

Chaffer CL, Thomas DM, Thompson EW, Williams ED. PPARγ-independent induction of growth arrest and apoptosis in prostate and bladder carcinoma. *BMC Cancer* 2006; 6:53

Chang TH, Szabo E. Enhanced growth inhibition by combination differentiation therapy with ligands of peroxisome proliferator-activated receptor-gamma and inhibitors of histone deacetylase in adenocarcinoma of the lung. *Clin Cancer Res.* 2002; 8(4):1206-12

Chang NW, Tsai MH, Lin C, Hsu HT, Chu PY, Yeh CM, Chiu CF, Yeh KT. Fenofibrate exhibits a high potential to suppress the formation *Biochim Biophys Acta.* 2010; [Epub ahead of print]

Chawla A, Barak Y, Nagy L, Liao D, Tontonoz P, Evans RM. PPAR-gamma dependent and independent effects on macrophage-gene expression *Nat Med.* 2001; 7(1):48-52

Chen GG, Lee JF, Wang SH, Chan UP, Ip PC, Lau WY. Apoptosis induced by activation of peroxisome-proliferator activated receptor-gamma is associated with Bcl-2 and NF-kappaB in human colon cancer. *Life Sci.* 2002; 70:2631–46

Chen YX, Zhong XY, Qin YF, Bing W, He LZ. 15d-PGJ2 inhibits cell growth and induces apoptosis of MCG-803 human gastric cancer cell line. *World J Gastroenterol.* 2003; 9(10):2149-53

Chen WC, Lin MS, Bai X. Induction of apoptosis in colorectal cancer cells by peroxisome proliferators-activated receptor gamma activation up-regulating PTEN and inhibiting PI3K activity. *Chin Med J (Engl).* 2005; 118:1477-81

Chen LL, Zhang JY, Wang BP. Renoprotective effects of fenofibrate in diabetic rats are achieved by suppressing kidney plasminogen activator inhibitor-1. *Vascul. Pharmacol.* 2006; 4:309–15

Chinetti G, Griglio S, Antonucci M, Torra IP, Delerive P, Majd Z, Fruchart JC, Chapman J, Najib J, Staels B. Activation of proliferator-activated receptors *J Biol Chem.* 1998;273(40):25573-80

Chinetti G, Lestavel S, Bocher V, Remaley AT, Neve B, Torra IP, Teissier E, Minnich A, Jaye M, Duverger N, Brewer HB, Fruchart JC, Clavey V, Staels B. PPAR-alpha and PPAR-gamma activators induce cholesterol *Nat Med.* 2001; 7(1):53-8

Clark RB, Bishop-Bailey D, Estrada-Hernandez T, Hla T, Puddington L, Padula SJ. The nuclear receptor PPAR-γ and immunoregulation: PPAR-γ mediates inhibition of helper T cell responses. *J Immunol.* 2000; 164:1364-71

Conway JG, Tomaszewski KE, Olson MJ, Cattley RC, Marsman DS, Popp JA. Relationship of oxidative damage to the hepatocarcinogenicity of the peroxisome proliferators di(2-ethylhexyl)phthalate and Wy 14,643. *Carcinogenesis.* 1989; 10:513-9

Cooper K, Squires H, Carroll C, Papaioannou D, Booth A, Logan RF, Maguire C, Hind D, Tappenden P. Chemoprevention of colorectal cancer *Health Technol Assess.* 2010; 14(32):1-206

Crisafulli C, Cuzzocrea S. The role of endogenous and exogenous ligands for the peroxisome proliferator-activated receptor α (PPAR-α) in the regulation of inflammation in macrophages. *Shock* 2009; 32:62-73

Cronet P, Petersen JF, Folmer R, Blomberg N, Sjöblom K, Karlsson U, Lindstedt EL, Bamberg K. Structure of the PPARalpha and -gamma ligand binding domain in complex with AZ 242; ligand selectivity and agonist activation in the PPAR family. *Structure.* 2001; 9(8):699-706

Crosby MB, Zhang J, Nowling TM, Svenson JL, Nicol CJ, Gonzalez FJ, Gilkeson GS. Inflammatory modulation of PPAR gamma expression and activity. *Clin Immunol* .2006; 118:276-83

Cunard R, Ricote M, Dicampli D, Archer DC, Kahn DA, Glass CK, Kelly CJ. Regulation of cytokine expression by ligands of peroxisome proliferator activated receptors. *J Immunol.* 2002; 168:2795-802

Dahten A, Koch C, Ernst D, ller CS, Hartmann S,, Worm M. Systemic PPARγ ligation inhibits allergic immune response in the skin. *J Invest Dermatol.* 2008; 128:2211–8

Debril MB, Renaud JP, Fajas L, Auwerx J. The pleiotropic functions of peroxisome proliferator-activated receptor γ. *J Mol Med.* 2001; 79:30-47

Deeb SS, Fajas L, Nemoto M, Pihlajamaki J, Mykkanen L, Kuusisto J, Laakso M, Fujimoto W, Auwerx J. A Pro12Ala substitution in PPARg2 associated with decreased receptor activity, lower body mass index and improved insulin sensitivity. *Nat Genet.* 1998; 20:284-7

Defaux A, Zurich MJ, Braissant O, Honegger P, Monnet-Tschudi F. Effects of the PPAR-β agonist GW501516 in an in vitro model of brain inflammation and antibody-induced demyelination. *J Neuroinflammation.* 2009; 6:15

Delerive P, Gervois P, Fruchart JC, Staels B. Induction of IkappaBalpha expression as a mechanism contributing to the anti-inflammatory activities of peroxisome proliferator-activated receptor-alpha activators. *J Biol Chem* 2000;275:36703–7

Delerive P, J-C Fruchart J-C, Staels B. Peroxisome proliferator-activated receptors in inflammation control. *J Endocrinol.* 2001; 169:453–9

Demerjian M, Man MQ, Choi EH, Brown BE, Crumrine D, Chang S, Mauro T, Elias PM, Feingold KR. Topical treatment with thiazolidinediones, activators of peroxisome proliferator-activated receptor-gamma, normalizes epidermal homeostasis in a murine hyperproliferative disease model. *Exp Dermatol.* 2006; 15(3):154-60

Desvergne B, Wahli W. Peroxisome proliferator-activated receptors: nuclear control of metabolism. *Endocr Rev.* 1999; 20(5):649-88

Devchand PR, Keller H, Peters JM, Vazquez M, Gonzalez FJ, Wahli W. The PPAR□-leukotriene B4 pathway to inflammation control. *Nature.* 1996; 384:39-43

Din FV, Theodoratou E, Farrington SM, Tenesa A, Barnetson RA, Cetnarskyj R, Stark L, Porteous ME, Campbell H, Dunlop MG. Effect of aspirin and NSAIDs *Gut.* 2010; 59(12):1670-9

Dongxia L, Qi H, Lisong L, Jincheng G. Association of peroxisome proliferator-activated receptorgamma gene Pro12Ala and C161T polymorphisms with metabolic syndrome. *Circ J.* 2008;72(4):551-7

Douglas JA, Erdos MR, Watanabe RM, Braun A, Johnston CL, Oeth P, Mohlke KL, Valle TT, Ehnholm C, Buchanan TA, Bergman RN, Collins FS, Boehnke

M, Tuomilehto J. The peroxisome proliferator-activated receptor-γ2 Pro12Ala variant: association with type 2 diabetes and trait differences. *Diabetes* 2001; 50:886-90

Dreyer C, Krey G, Keller H, Givel F, Helftenbein G, Wahli W. Control of the peroxisomal beta-oxidation pathway by a novel family of nuclear hormone receptors. *Cell.* 1992; 68(5):879-87

Dunn SE, Bhat R, Straus DS, Sobel RA, Axtell R, Johnson A, Nguyen K, Mukundan L, Moshkova M, Dugas JC, Chawla A, Steinman L. Peroxisome proliferator-activated receptor delta limits the expansion of pathogenic Th cells during central nervous system autoimmunity. *J Exp Med.* 2010; 207(8):1599-608

Durbin RJ. Thiazolidinedione therapy *Diabetes Obes Metab*. 2004; 6(4):280-5

Elstner E, Muller C, Koshizuka K, Williamson EA, Park D, Asou H, Shintaku P, Said JW, Heber D, Koeffler HP. Ligands for peroxisome proliferator-activated receptorgamma and retinoic acid receptor inhibit growth and induce apoptosis of human breast cancer cells in vitro and in BNX mice. *Proc. Natl. Acad. Sci. U.S.A*. 1998; 95:8806–11

Etgen GJ, Oldham BA, Johnson WT, Broderick CL, Montrose CR, Brozinick JT, Misener EA, Bean JS, Bensch WR, Brooks DA, Shuker AJ, Rito CJ, McCarthy JR, Ardecky RJ, Tyhonas JS, Dana SL, Bilakovics JM, Paterniti JR Jr, Ogilvie KM, Liu S, Kauffman RF. A tailored therapy for the metabolic syndrome: the dual peroxisome proliferator-activated receptor-alpha/gamma agonist LY465608 ameliorates insulin resistance and diabetic hyperglycemia while improving cardiovascular risk factors in preclinical models. *Diabetes*. 2002; 51(4):1083-7

Fajas L, Auboeuf D, Raspé E, Schoonjans K, Lefebvre AM, Saladin R, Najib J, Laville M, Fruchart JC, Deeb S, Vidal-Puig A, Flier J, Briggs MR, Staels B, Vidal H, Auwerx J. The organization, promoter analysis, and expression of the human PPARgamma gene. *J Biol Chem.* 1997; 272(30):18779-89

Fajas L, Fruchart JC, Auwerx J. PPARgamma3 mRNA: a distinct PPARgamma mRNA subtype transcribed from an independent promoter. *FEBS Lett.* 1998; 438(1-2):55-60

Fischer B, von Knethen A, Brune B. Dualism of oxidized lipoproteins in provoking and attenuating the oxidative burst in macrophages: role of peroxisome proliferatoractivated receptor-γ. *J Immunol.* 2002; 168(6):2828–34

Flanders KC, Burmester JK. Medical applications of transforming growth factor-beta. *Clin. Med. Res.* 2003; 1:13–20

Floyd ZE, Stephens JM. Interferon-gamma-mediated activation and ubiquitin-proteasome-dependent degradation of PPARgamma in adipocytes. *J Biol Chem*. 2002; 277:4062–8

Foreman JE, Sorg JM, McGinnis KS, Rigas B, Williams JL, Clapper ML, Gonzalez FJ, Peters JM. Regulation of peroxisome proliferator-activated receptor-beta/delta by the APC/beta-CATENIN pathway and nonsteroidal antiinflammatory drugs. *Mol Carcinog.* 2009; 48(10):942-52

Forman BM, Tontonoz P, Chen J, Brun RP, Spiegelman BM, Evans RM. 15-Deoxy-delta 12, 14-prostaglandin J2 is a ligand for the adipocyte determination factor PPAR gamma. *Cell.* 1995; 83(5):803-12

Fredenrich A, Grimaldi PA. PPAR delta: an uncompletely known nuclear receptor. *Diabetes Metab.* 2005; 31(1):23-7

Frederiksen L, Brodbaek K, Fenger M, Jorgensen T, Borch-Johnsen K,. Madsbad S, Urhammer SA. Comment: studies of the Pro12Ala polymorphism of the PPAR-gamma gene in the Danish MONICA cohort: homozygosity of the Ala allele confers a decreased risk of the insulin resistance syndrome. *J Clin Endocrinol Metab*. 2001; 87:3989–92

Fujiwara T, Horikoshi H. Troglitazone and related compounds *Life Sci.* 2000; 67(20):2405-16

Fukumoto K, Yano Y, Virgona N, Hagiwara H, Sato H, Senba H, Suzuki K, Asano R, Yamada K, Yano T. Peroxisome proliferator-activated receptor delta as a molecular target to regulate lung cancer cell growth. *FEBS Lett* 2005; 79:3829-36

Gbaguidi FG, Chinetti G, Milosavljevic D, Teissier E, Chapman J, Olivecrona G, Fruchart JC, Griglio S, Fruchart-Najib J, Staels B. Peroxisome proliferator-activated receptor (PPAR) agonists decrease lipoprotein lipase secretion *FEBS Lett*. 2002; 512(1-3):85-90

Genini D, Catapano CV. Block of nuclear receptor ubiquitination. A mechanism of ligand-dependent control of peroxisome proliferator-activated receptor delta activity. *J Biol Chem.* 2007; 282(16):11776-85

Genini D, Carbone GM, Catapano CV. Multiple Interactions between Peroxisome Proliferators-Activated Receptors and the Ubiquitin-Proteasome System and Implications for Cancer Pathogenesis. *PPAR Res.* 2008; 2008:195065 (11 pages)

Gervois P, Chopin-Delannoy S, Fadel A, Dubois S, Kosykh V, Fruchart JC, Najïb J, Laudet V, Staels B. Fibrates increase human REV-ERBalpha expression in liver via a novel peroxisome proliferator-activated receptor response element. *Mol Endocrinol*. 1999; 13(3):400–9

Ghoussaini M, Meyre D, Lobbens S, Charpentier G, Clément K, Charles MA, Tauber M, Weill J, Froguel P. Implication of the Pro12Ala polymorphism *BMC Med Genet.* 2005; 6:11

Girroir EE, Hollingshead HE, Billin AN, Willson TM, Robertson GP, Sharma AK, Amin S, Gonzalez FJ, Peters JM. Peroxisome proliferator-activated receptor-β/δ (PPARβ/δ) ligands inhibit growth of UACC903 and MCF7 human cancer cell lines. *Toxicology*. 2008; 243: 236–43

Göke R, Göke A, Göke B, El-Deiry WS, Chen Y. Pioglitazone inhibits growth *Digestion*. 2001; 64(2):75-80

Gocke AR, Hussain RZ, Yang Y, Peng H, Weiner J, Ben LH, Drew PD, Stuve O, Lovett-Racke AE, Racke MK. Transcriptional modulation of the immune response by peroxisome proliferator-activated receptor-α agonists in autoimmune disease. *J Immunol*. 2009;182:4479-87

González Sánchez JL, Serrano Ríos M, Fernández Perez C, Laakso M, Martínez Larrad MT. Effect of the Pro12Ala polymorphism of the peroxisome proliferator-activated receptor gamma-2 gene on adiposity, insulin sensitivity and lipid profile in the Spanish population. *Eur J Endocrinol.* 2002; 147(4):495-501

Goyenechea E, Parra MD, Martınez JA. Weight regain after slimming induced by an energy-restricted diet depends on interleukin-6 and peroxisome-proliferatoractivated-receptor-γ2 gene polymorphisms. *Br J Nutrition*. 2006; 96(5):965-72

Grabacka M, Reiss K. Anticancer Properties of PPARalpha-Effects on Cellular Metabolism and Inflammation. *PPAR Res*. 2008; 2008:930705

Gray JP, Burns KA, Leas TL, Perdew GH, Vanden Heuvel JP. Regulation of peroxisome proliferatoractivated receptor alpha by protein kinase C. *Biochemistry*. 2005; 44:10313–21

Green S, Tugwood JD, Issemann I. The molecular mechanism of peroxisome proliferator action: a model for species differences and mechanistic risk assessment. *Toxicol Lett*. 1992; 64/65:131-9

Green S, Issemann I, Tugwood JD. The molecular mechanism of peroxisome proliferator action in "*Peroxisomes. Biology and importance in toxicology and medicine*" Gibson G. and Lake B. (eds), Taylor and Francis Inc., London, 1993; Chapter 4 pp. 99-118.

Greene ME, Blumberg B, McBride OW, Yi HF, Kronquist K, Kwan K, Hsieh L, Greene G, Nimer SD. Isolation of the human peroxisome proliferator activated receptor gamma cDNA: expression in hematopoietic cells and chromosomal mapping. *Gene Expr.* 1995; 4(4-5):281-99

Grimaldi PA. Regulatory functions of PPARβ in metabolism: implications for the treatment of metabolic syndrome. *Biochim Biophys Acta*. 2007; 1771:983–90

Guellich A, Damy T, Lecarpentier Y, Conti M, Claes V, Samuel JL, Quillard J, Hébert JL, Pineau T, Coirault C. Role of oxidative stress *Am J Physiol Heart Circ Physiol*. 2007; 293(1):H93-H102

Guri AJ, Mohapatra SK, Horne II WT, Hontecillas R, Bassaganya-Riera J. The role of T cell PPAR γ in mice with experimental inflammatory bowel disease. *BMC Gastroenterology* 2010; 10:60

Hamblin M, Chang L, Zhang J, Chen YE. The role of peroxisome proliferator-activated receptor gamma in blood pressure regulation. *Curr Hypertens Rep*. 2009; 11(4):239-45

Hansen JB, Zhang H, Rasmussen TH, Petersen RK, Flindt EN, Kristiansen K. Peroxisome proliferator-activated receptor delta (PPARdelta)- mediated regulation of preadipocyte proliferation and gene expression is dependent on cAMP signaling. *J Biol Chem*. 2001; 276:3175-82

Hara K, Okada T, Tobe K, Yasuda K, Mori Y, Kadowaki H, Hagura R, Akanuma Y, Kimura S, Ito C, Kadowaki T. The pro12ala polymorphism in PPAR γ may confer resistance to type 2 diabetes. *Biochem Biophys Res Commun*. 2000; 271:212–6

Harman FS, Nicol CJ, Marin HE, Ward JM, Gonzalez FJ, Peters JM. Peroxisome proliferator-activated receptor-delta attenuates colon carcinogenesis. *Nat Med*. 2004; 10:481-3

Harris SG, Phipps RP. Prostaglandin D2, its metabolite 15-d-PGJ2, and peroxisome proliferator activated receptor-γ agonists induce apoptosis in transformed, but not normal, human T lineage cells. *Immunology*. 2002; 105:23-34

Harris SG, Padilla J, Koumas L, Denise R, Phipps RP. Prostaglandin as modulators of immunity. *Trends Immunol*. 2002; 23:144-50

Hasegawa H, Takano H, Komuro I. therapeutic implications of PPARγ in cardiovascular diseases. *PPAR Res*. 2010. doi:10.1155/2010/876049

Hashizume S, Akaike M, Azuma H, Ishikawa K, Yoshida S, Sumitomo-Ueda Y, Yagi S, Ikeda Y, Iwase T, Aihara KI, Abe M, Sata M, Matsumoto T. Activation of Peroxisome Proliferator-activated Receptor α in Megakaryocytes Reduces Platelet-derived Growth Factor-BB in Platelets. *J Atheroscler Thromb*. 2010 Nov 2. [Epub ahead of print]

Hauser S, Adelmant G, Sarraf P, Wright HM, Mueller E, Spielgelman BM. Degradation of the peroxisome proliferator-activated receptor-γ is linked to ligand dependent activation. *J Biol Chem*. 2000; 275:18527-33

He TC, Chan TA, Vogelstein B, Kinzler KW. PPARδ is an APC-regulated target of nonsteroidal anti-inflammatory drugs. *Cell* 1999; 99:335–45

Heneka MT, Landreth GE. PPARs in the brain. Biochim Biophys Acta. 2007; 1771:1031–45

Hodges M, Tissot C, Freemont PS. Protein regulation: tag wrestling with relatives of ubiquitin. *Curr. Biol.* 1998; 8(21):R749–52

Hostetler HA, Petrescu AD, Kier AB, Schroeder F. Peroxisome proliferator-activated receptor α interacts with high affinity and is conformationally responsive to endogenous ligands *J Biol Chem*. 2005; 280:18667-82

Hotta K, Gustafson TA, Yoshioka S, Ortmeyer HK, Bodkin NL, Hansen BC. Relationships of PPARgamma and PPARgamma2 mRNA levels to obesity, diabetes and hyperinsulinaemia in rhesus monkeys. *Int J Obes Relat Metab Disord.* 1998; 22(10):1000-10

Houseknecht KL, Bidwell CA, Portocarrero CP, Spurlock ME. Expression and cDNA cloning of porcine peroxisome proliferator-activated receptor gamma (PPARgamma). *Gene*. 1998; 225(1-2):89-96

Hu E, Kim JB, Sarraf P, Spiegelman BM. Inhibition of adipogenesis through MAP kinase mediated phosphorylation of PPAR-γ. *Science*. 1996; 274:2100–5

Huguenin GV, Rosa G. The Ala allele in the PPAR-gamma2 gene is associated with reduced risk of type 2 diabetes mellitus in Caucasians and improved insulin sensitivity in overweight subjects. *Br J Nutr.* 2010; 104(4):488-97

Ijpenberg A, Jeannin E, Wahli W, Desvergne B. Polarity and specific sequence requirements of peroxisome proliferator-activated receptor (PPAR)/retinoid X receptor heterodimer binding to DNA. A functional analysis of the malic enzyme gene PPAR response element. *J Biol Chem*. 1997; 272(32):20108−17

Isseman I, Green S. Activation of a number of the steroid receptor superfamily by peroxisome proliferators. *Nature*. 1990; 347:645–50

Iwamoto Y, Kuzuya T, Matsuda A, Awata T, Kumakura S, Inooka G, Shiraishi I. Effect of new oral antidiabetic agent CS-045 on glucose *Diabetes Care.* 1991; 14(11):1083-6

Jay MA, Ren J. Peroxisome proliferator-activated receptor (PPAR) in metabolic syndrome and type 2 diabetes mellitus. *Curr Diabetes Rev*. 2007; 3(1):33-9

Jiang C, Ting AT, Seed B. PPAR-γ agonist inhibit production of monocyte inflamatory cytokines. *Nature*. 1998; 391:82-6

Kaipainen A, Kieran MW, Huang S, Butterfield C, Bielenberg D, Mostoslavsky G, Mulligan R, Folkman J, Panigrahy D. PPARalpha deficiency in inflammatory cells suppresses tumor growth. *PLoS One*. 2007; 2(2):e260

Karin M, Ben-Neriah Y. Phosphorylation meets ubiquitination: the control of NF-κB activity. *Annu Rev Immunol.* 2000; 18:621-63

Kasai H, Okadi Y, Nishimura S, Rao MS, Reddy JK. Formation of 8-hydroxydeoxyguanosine in liver DNA of rats following long-term exposure to a peroxisome proliferator. *Cancer Res.* 1989; 49: 2603-5

Keller H, Givel F, Perroud M, Wahli W. Signaling cross-talk between peroxisome proliferator-activated receptor/retinoid X receptor and estrogen receptor through estrogen response elements. *J Mol Endocrinol.* 1995; 19:794-804

Kersten S, Seydoux J, Peters JM, Gonzalez FJ, Desvergne B, Wahli W. Peroxisome proliferator-activated receptor alpha mediates the adaptive response *J Clin Invest.* 1999; 103(11):1489-98

Kim DJ, Bility MT, Billin AN, Willson TM, Gonzalez FJ, Peters JM. PPARβ/δ selectively induces differentiation and inhibits cell proliferation. *Cell Death Differ.* 2006; 13:53–60

Kim HJ, Kim JY, Meng Z, Wang LH, Liu F, Conrads TP, Burke TR, Veenstra TD, Farrar WL. 15-deoxy-$\Delta^{12,14}$-prostaglandin J_2 inhibits transcriptional activity of estrogen receptor-α via covalent modification of DNA-binding domain. *Cancer Res.* 2007; 67:2595-602

Kitamura S, Miyazaki Y, Shinomura Y, Kondo S, Kanayama S, Matsuzawa Y. Peroxisome proliferator-activated receptor gamma induces growth *Jpn J Cancer Res.* 1999; 90(1):75-80

Kliewer SA, Forman BM, Blumberg B, Ong ES, Borgmeyer U, Mangelsdorf DJ, Umesono K, Evans RM. Differential expression and activation of a family of murine peroxisome proliferator-activated receptors. *Proc Natl Acad Sci U S A.* 1994; 91:7355–9

Kliewer SA, Sundseth SS, Jones SA, Brown PJ, Wisely GB, Koble CS, Devchand P, Wahli W, Willson TM, Lenhard JM, Lehmann JM. Fatty acids and eicosanoids regulate gene expression through direct interactions with peroxisome proliferator-activated receptors alpha and gamma. *Proc. Natl. Acad. Sci. U.S.A.* 1997; 94:4318–23

Koeffler HP. Peroxisome proliferator-activated receptor gamma and cancers. *Clin Cancer Res.* 2003; 9(1):1-9

Kopelovich L, Fay JR, Glazer RI, Crowell JA. Peroxisome proliferator-activated receptor modulators as potential chemopreventive agents. *Mol Cancer Ther.* 2002; 1:357-63

Kota BP, Huang TH, Roufogalis BD. An overview on biological mechanisms of PPARs. *Pharmacol Res.* 2005; 51:85-94

Kraupp-Grasl B, Huber W, Putz B, Gerbracht U, Schulte-Hermann R. Tumor promotion by the peroxisome proliferator nafenopin involving a specific subtype of altered foci in rat liver. *Cancer Res.* 1990; 50:3701-8

Krogsdam AM, Nielsen CA, Neve S, Holst D, Helledie T, Thomsen B, Bendixen C, Mandrup S, Kristiansen K. Nuclear receptor corepressor-dependent repression of peroxisome-proliferator-activated receptor delta-mediated transactivation. *Biochem J.* 2002; 363(Pt 1):157-65

Kwak BR, Myit S, Mulhaupt F, Veillard N, Rufer N, Roosnek E, Mach F. PPARγ but not PPARα ligands are potent repressors of major histocompatibility complex class II induction in atheroma-associated cells. *Circ Res.* 2002; 90:356-62

Lanas A. Nonsteroidal antiinflammatory drugs and cyclooxygenase inhibition in the gastrointestinal tract: a trip from peptic ulcer to colon cancer. *Am J Med Sci.* 2009; 338(2):96-106

Larsen TM, Toubro S, Astrup A. PPARgamma agonists in the treatment of type II diabetes: is increased fatness commensurate with long-term efficacy? *Int. J. Obes. Relat. Metab. Disord.* 2003; 27:147-61

Lazar MA. PPAR gamma, 10 years later. *Biochimie.* 2005; 87:9-13

Lazennec G, Canaple L, Saugy D, Wahli W. Activation of peroxisome proliferator-activated receptors (PPARs) by their ligands and protein kinase A activators. *Mol Endocrinol* 2000; 14:1962-75

Lee H, Shi W, Tontonoz P, Wang S, Subbanagounder G, Hedrick CC, Hama S, Borromeo C, Evans RM, Berliner JA, Nagy L. Role for peroxisome proliferator-activated receptor alpha in oxidized phospholipid-induced synthesis *Circ Res.* 2000; 87(6):516-21

Lee CH, Olson P, Hevener A, Mehl I, Chong LW, Olefsky JM, Gonzalez FJ, Ham J, Kang H, Peters JM, Evans RM. PPARdelta regulates glucose metabolism and insulin sensitivity. *Proc Natl Acad Sci U S A.* 2006; 103(9):3444-9

Lefebvre P, Chinetti G, Fruchart J-C, Staels B. Sorting out the roles of PPARα in energy metabolism and vascular homeostasis. *J Clin Invest.* 2006; 116: 571–80

Lemay DG, Hwang DH. Genome-wide identification of peroxisome proliferator response elements using integrated computational genomics. *J Lipid Res.* 2006; 47(7):1583-7

Leonardini A, Laviola L, Perrini S, Natalicchio A, Giorgino F. Cross-talk between PPARγ and insulin signaling and modulation of insulin sensitivity. *PPAR Res.* 2009; 2009:818945 (12 pages) doi:10.1155/2009/818945

Li M, Lee TW, Mok TS, Warner TD, Yim AP, Chen GG. Activation of peroxisome proliferator-activated receptor-gamma by troglitazone (TGZ) inhibits human*J Cell Biochem*. 2005; 1;96(4):760-74

Liao SY, Zeng ZR, Leung WK, Zhou SZ, Chen B, Sung JJ, Hu PJ. Peroxisome proliferator-activated receptor-gamma Pro12Ala polymorphism, *Helicobacter pylori* infection and non-cardia gastric carcinoma in Chinese. *Aliment Pharmacol Ther*. 2006; 23(2):289-94

Liou JY, Ghelani D, Yeh S, Wu KK. Nonsteroidal anti-inflammatory drugs induce colorectal cancer cell apoptosis by suppressing 14-3-3ε. *Cancer Res*. 2007; 67:3185–91

Lu YL, Li GL, Huang HL, Zhong J, Dai LC. Peroxisome proliferator-activated receptor-gamma 34C>G polymorphism and colorectal cancer risk: a meta-analysis. *World J Gastroenterol*. 2010; 16(17):2170-5

Lytle C, Tod TJ, Vo KT, Lee JW, Atkinson RD, Straus DS. The peroxisome proliferator-activated receptor gamma ligand rosiglitazone delays the onset of inflammatory bowel disease in mice with interleukin 10 deficiency. *Inflamm Bowel Dis*. 2005; 11:231-43

Maggiora M, Oraldi M, Muzio G, Canuto RA. Involvement of PPARα and PPARγ in apoptosis *Cell Biochem Funct*. 2010; 28(7):571-7

Majai G, Sarang Z, Csomos K, Zahuczky G, Fesus L. PPARgamma-dependent regulation of human macrophages in phagocytosis of apoptotic cells. *Eur J Immunol*. 2007; 37:1343-54

Mancini FP,Vaccaro O, Sabatino L,Tufano A, RivelleseAA, Riccardi G, Colantuoni V. Pro12Ala substitution in the peroxisome proliferator activated receptor-γ2 is not associated with type 2 diabetes. *Diabetes*. 1999; 48:1466–8

Mangelsdorf D.J. and Evans R.M. The RXR heterodimers and orphan receptors. *Cell*. 1995; 83: 841-50

Marin HE, Peraza MA, Billin AN, Willson TM, Ward JM, Kennett MJ, Gonzalez FJ, Peters JM. Ligand activation of peroxisome proliferator-activated receptor β/δ (PPARβ/δ) inhibits colon carcinogenesis. *Cancer Res*. 2006; 66:4394–401

Marsman DS, Cattley RC, Conway JG, Popp JA. Relationship of hepatic peroxisome proliferation and replicative DNA synthesis to the hepatocarcinogenicity of the peroxisome proliferators di(2-ethylhexyl)phthalate and [4-chloro - 6 - (2,3-xylidino) - 2 - pyrimidinylthio] acetic acid (Wy 14,643) in rats. *Cancer Res*. 1988; 48:6739-44

Marx N, Schonbeck U, Lazar MA, Libby P, Plutzky J. Peroxisome proliferator-activated receptor gamma activators inhibit gene expression and migration in human vascular smooth muscle cells. *Circ Res* 1998; 83(11):1097-103

Marx N, Sukhova GK, Collins T, Libby P, Plutzky J. PPARα activators inhibit cytokine-induced vascular cell adhesion molecule-1 expression in human endothelial cells," *Circulation*. 1999; 99(24):3125–31

Mehta RG, Williamson E, Patel MK, Koeffler HP. A ligand of peroxisome proliferator-activated receptor gamma, retinoids, and prevention of preneoplastic mammary lesions. *J Natl Cancer Inst.* 2000; 92(5):418-23

Meirhaeghe A, Fajas L, Helbecque N, Cottel D, Lebel P, Dallongeville J, Deeb S, Auwerx J, Amouyel P. A genetic polymorphism of the peroxisome proliferator-activated receptor gamma gene influences plasma leptin levels in obese humans. *Hum Mol Genet.* 1998; 7:435–40

Meshkani R, Taghikhani M, Larijani B, Bahrami Y, Khatami S, Khoshbin E, Ghaemi A, Sadeghi S, Mirkhani F, Molapour A, Adeli K. Pro12Ala polymorphism of the peroxisome proliferator-activated receptor-gamma2 (PPARgamma-2) gene is associated with greater insulin sensitivity and decreased risk of type 2 diabetes in an Iranian population. *Clin Chem Lab Med.* 2007; 45(4):477-82

Michalik L, Auwerx J, Berger JP, Chatterjee VK, Glass CK, Gonzalez FJ, Grimaldi PA, Kadowaki T, Lazar MA, O'Rahilly S, Palmer CN, Plutzky J, Reddy JK, Spiegelman BM, Staels B, Wahli W. International Union of Pharmacology. LXI. Peroxisome proliferator-activated receptors. *Pharmacol Rev.* 2006; 58(4):726-41

Michalik L, Wahli W. PPARs mediate lipid signaling in inflammation and cancer. *PPAR Res.* 2008; 2008:134059

Miller JL. Insulin resistance *Postgrad Med.* 2003; Spec No:27-34

Moraes LA, Piqueras L, Bishop-Bailey D. Peroxisome proliferator-activated receptors and inflammation. *Pharmacol Ther.* 2006; 110(3):371–85

Mori H, Ikegami H, Kawaguchi Y, Seino S, Yokoi N, Takeda J, Inoue I, Seino Y, Yasuda K, Hanafusa T, Yamagata K, Awata T, Kadowaki T, Hara K, Yamada N, Gotoda T, Iwasaki N, Iwamoto Y, Sanke T, Nanjo K, Oka Y, Matsutani A, Maeda E, Kasuga M: The Pro12 →Ala substitution in PPAR-γ is associated with resistance to development of diabetes in the general population: possible involvement in impairment of insulin secretion in individuals with type 2 diabetes. *Diabetes*. 2001; 50:891–94

Motojima K. Peroxisome proliferator-activated receptor (PPAR): Structure, mechanism of activation and diverse functions. *Cell Struct Funct.* 1993; 18:267-77

Motomura W, Takahashi N, Nagamine M, Sawamukai M, Tanno S, Kohgo Y, Okumura T. Growth arrest by troglitazone is mediated by p27Kip1 accumulation, which results from dual inhibition of proteasome activity and

Skp2 expression in human hepatocellular carcinoma cells. *Int J Cancer*. 2004; 108(1):41-6

Mueller E, Sarraf P, Tontonoz P, Evans RM, Martin KJ, Zhang M, Fletcher C, Singer S, Spiegelman BM. Terminal differentiation of human *Mol Cell.* 1998; 1(3):465-70

Mukundan L, Odegaard JI, Morel CR, Heredia RE, Mwangi JW, Ricardo-Gonzalez, RR, Goh YPS, Eagle AR, Dunn SE, Awakuni JUH, Nguyen KD, Steinman L, Michie SA, Chawla A. PPAR-δ senses and orchestrates clearance of apoptotic cells to promote tolerance. *Nat Med*. 2009; 15(11):1266–72

Murtaugh MA, Ma KN, Caan BJ, Sweeney C, Wolff R, Samowitz WS, Potter JD, Slattery ML. Interactions of peroxisome proliferator-activated receptor {gamma} and diet *Cancer Epidemiol Biomarkers Prev*. 2005; 14(5):1224-9

Muzio G, Martinasso G, Trombetta A, Di Simone D, Canuto RA, Maggiora M. HMG-CoAreductase and PPARalpha are involved in clofibrate-induced apoptosis in human keratinocytes. *Apoptosis*. 2006; 11:265-75

Nagy L, Tontonoz P, Alvarez JG, Chen H, Evans RM. Oxidized LDL regulates macrophage gene expression through ligand activation of PPARgamma. *Cell.* 1998; 93:229-40

Narala VR, Adapala RK, Suresh MV, Brock TG, Peters-Golden M, Reddy RC. Leukotriene B4 is a physiologically relevant endogenous peroxisome proliferator-activated receptor-alpha agonist. *J Biol Chem*. 2010; 285(29):22067-74

Neve BP, Corseaux D, Chinetti G, Zawadzki C, Fruchart JC, Duriez P, Staels B, Jude B. PPARalpha agonists inhibit tissue factor expression in human monocytes and macrophages. *Circulation*. 2001; 103:207–12

Nguyen-Khoa, R., Z. Massy, V. Witko-Sarsat, S. Canteloup, M. Kebebe, B. Lacour, T. Drüecke, B. Descamps-Latscha. Oxidized low-density lipoprotein induces macrophage respiratory burst via its protein moiety: a novel pathway in atherogenesis?. *Biochem Biophys Res Commun. 1999; 263:804-9*

Nissen SE, Wolski K, Topol EJ. Effect of muraglitazar on death and major adverse cardiovascular events in patients with type 2 diabetes mellitus. *JAMA*. 2005; 294(20):2581-6

Odegaard JI, Ricardo-Gonzalez RR, Goforth MH, Morel CR, Subramanian V, Mukundan L, Eagle AR, Vats D, Brombacher F, Ferrante AW, Chawla A. Macrophage-specific PPARgamma controls alternative activation and improves insulin resistance. *Nature*. 2007; 447:1116–20

Ogawa S, Urano T, Hosoi T, Miyao M, Hoshino S, Fujita M, Shiraki M, Orimo H, Ouchi Y, Inoue S. Association of bone mineral density with a polymorphism of the peroxisome proliferator-activated receptor gamma gene: PPARgamma expression in osteoblasts. *Biochem Biophys Res Commun.* 1999; 260:122–6

Orasanu G, Ziouzenkova O, Devchand PR, Nehra V, Hamdy O, Horton ES, Plutzky J. The peroxisome proliferator-activated receptor-γ agonist pioglitazone represses inflammation in a peroxisome proliferator-activated receptor-α–dependent manner in vitro and in vivo in mice. *J Am Coll Cardiol* 2008; 52(10):869-81

Oyama T, Toyota, Waku T, Hirakawa Y, Nagasawa N, Kasuga JI, Hashimoto Y, Miyachi H, Morikawa K. Adaptability and selectivity of human peroxisome proliferator-activated receptor (PPAR) pan agonists revealed from crystal structures. *Acta Crystallogr D Biol Crystallogr.* 2009; 65(8):786-95

Panigrahy D, Kaipainen A, Huang S, Butterfield CE, Barnés CM, Fannon M, Laforme AM, Chaponis DM, Folkman J, Kieran MW. PPARalpha agonist *Proc Natl Acad Sci U S A.* 2008; 105(3):985-90

Park BH, Vogelstein B, Kinzler KW. Genetic disruption of PPAR δ decreases the tumorigenicity of human colon cancer cells. *Proc Natl Acad Sci U S A.* 2001; 98:2598–603

Park CW, Kim HW, Ko SH, Chung HW, Lim SW, Yang CW, Chang YS, Sugawara A, Guan Y, Breyer MD. Accelerated diabetic nephropathy in mice lacking the peroxisome proliferator-activated receptor-α. *Diabetes.* 2006a 55:885–93

Park CW, Zhang Y, Zhang X, Wu Y, Chen L, Cha DR, Su D, Hwang MT, Fan X, Davis L, Striker G, Zheng F, Breyer M, Guan Y. 2006b. PPAR□ agonist fenofibrate improves diabetic nephropathy in db/db mice. *Kidney Int.* 2006b; 69:1511–7

Pasceri V, Wu HD, Willerson JT, Yeh ETH. Modulation of vascular inflammation in vitro and in vivo by peroxisome proliferator-activated receptor-γ activators. *Circulation* 2000, 101(3):235-8

Patsouris D, Mandard S, Voshol PJ, Escher P, Tan NS, Havekes LM, Koenig W, März W, Tafuri S, Wahli W, Müller M, Kersten S. PPARalpha governs glycerol *J Clin Invest.* 2004; 114(1):94-103

Pecqueur C, Bui T, Gelly C, Hauchard J, Barbot C, Bouillaud F, Ricquier D, Miroux B, Thompson CB. Uncoupling protein-2 controls proliferation *FASEB J.* 2008; 22(1):9-18

Perissi V, Rosenfeld MG. Controlling nuclear receptors: the circular logic of cofactor cycles. *Nat Rev Mol Cell Biol.* 2005; 6(7):542–54

Peters JM, Cheung C, Gonzalez FJ. Peroxisome proliferator-activated receptor-α and liver cancer: where do we stand? *J Mol Med.* 2005; 83(10):774–85

Peters JM, Hollingshead HE, Gonzalez FJ. Role of peroxisome-proliferator-activated receptor beta/delta (PPARbeta/delta) in gastrointestinal tract function and disease. *Clin Sci (Lond).* 2008; 115(4):107-27

Peters JM, Gonzalez FJ. Sorting out the functional role(s) of peroxisome proliferator-activated receptor-β/δ (PPARβ/δ) in cell proliferation and cancer. *Biochimica et Biophysica Acta*. 2009; 1796:230–41

Planavila A, Rodriguez-Calvo R, Jove M, Michalik L, Wahli W, Laguna JC, Vazquez-Carrera M. Peroxisome proliferator-activated receptor beta/delta activation inhibits hypertrophy in neonatal rat cardiomyocytes. *Cardiovasc. Res*. 2005; 65:832–41

Polak PE, Kalinin S, Dello Russo C, Gavrilyuk V, Sharp A, Peters JM, Richardson J, Willson TM, Weinberg G, Feinstein DL. Protective effects of a peroxisome proliferator-activated receptor-beta/delta agonist in experimental autoimmune encephalomyelitis. *J Neuroimmunol.* 2005; 168(1-2):65-75

Poulsen P, Andersen G, Fenger M, Hansen T, Echwald SM, Volund A, Beck-Nielsen H, Pedersen O, Vaag A. Impact of two common polymorphisms in the PPARgamma gene on glucose tolerance and plasma insulin profiles in monozygotic and dizygotic twins: thrifty genotype, thrifty phenotype, or both? *Diabetes*. 2003; 52:194–8

Pyper SR, Viswakarma N, Yu S, Reddy JK. PPARalpha: energy *Nucl Recept Signal*. 2010; 8:e002

Qin C, Burghardt R, Smith R, Wormke M, Stewart J, Safe S. Peroxisome proliferator-activated receptor γ agonists induce proteasome-dependent degradation of cyclin D1 and estrogen receptor α in MCF-7 breast cancer cells. *Cancer Res*. 2003; 63(5):958–64

Qu A, Shah YM, Matsubara T, Yang Q, Gonzalez FJ. PPARalpha-dependent activation of cell cycle *Toxicol Sci.* 2010; 118(2):404-10

Quinn CE, Hamilton PK, Lockhart CJ, McVeigh GE. Thiazolidinediones: effects on insulin resistance and the cardiovascular system. *Br J Pharmacol.* 2008; 153(4):636-45

Rangwala SM, Rhoades B, Shapiro JS, Rich AS, Kim JK, Shulman GI, Kaestner KH, Lazar MA. Genetic modulation of PPARgamma phosphorylation regulates insulin sensitivity. *Dev Cell.* 2003; 5:657–63

Rao AS, Reddy JK. Hepatocarcinogenesis of peroxisome proliferators. *Ann. N.Y. Acad. Sci.* 1996; 804:573-87

Reddy JK, Chu R. Peroxisome proliferator-induced pleiotropic responses: pursuit of a phenomenon. *Ann. N.Y. Acad. Sci.* 1996; 804:176-201

Reddy JK, Lalwani ND. Carcinogenesis by hepatic peroxisome proliferators: evaluation of the risk of hypolipedemic drugs an industrial plasticizers to humans. *CRC Crit Rev Toxicol.* 1983; 12:1-58

Rhee EJ, Kwon CH, Lee WY, Kim SY, Jung CH, Kim BJ, Sung KC, Kim BS, Oh KW, Kang JH, Park SW, Kim SW, Lee MH, Park JR. No association of Pro12Ala polymorphism *Circ J.* 2007; 71(3):338-42

Ricote M, Li AC, Willson TM, Kelly CJ, Glass CK. (1998). The peroxisome proliferator activator receptor-γ is a negative regulator of macrophage activation. *Nature* 1998; 391:79-82

Ricote M, Glass CK. PPARs and molecular mechanisms of transrepression. *Biochim Biophys Acta.* 2007; 1771(8):926-35

Ridker PM, Cook NR, Cheng S, Erlich HA, Lindpaintner K, Plutzky J, Zee RYL. Alanine for proline substitution in the peroxisome proliferator–activated receptor gamma-2 (PPARG2) gene and the risk of incident myocardial infarction. *Arterioscler Thromb Vasc Biol.* 2003; 23:859-63

Ristow M, Müller-Wieland D, Pfeiffer A, Krone W, Kahn CR. Obesity associated with a mutation *N Engl J Med.* 1998; 339(14):953-9

Roberts-Thomson SJ. Peroxisome proliferator-activated receptors *Immunol Cell Biol.* 2000; 78(4):436-41

Rohn TT, Wong SM, Cotman CW, Cribbs DH. 15-deoxy-delta12,14-prostaglandin J2, a specific ligand for peroxisome proliferator-activated receptor-gamma, induces neuronal apoptosis. *Neuroreport.* 2001; 12(4):839-43

Romanowska M, Al Yacoub N, Seidel H, Donandt S, Gerken H, Phillip S, Haritonova N, Artuc M, Schweiger S, Sterry W, Foerster J. PPARδ enhances keratinocyte proliferation in psoriasis and induces heparin-binding EGF-like growth factor. *J Invest Dermatol.* 2008; 128:110–24

Rosen ED, Hsu CH, Wang X, Sakai S, Freeman MW, Gonzalez FJ, Spiegelman BM. C/EBPalpha induces adipogenesis through PPARgamma: a unified pathway. *Genes Dev.* 2002; 16:22-6

Rosen ED, Spiegelman BM. PPAR□: a nuclear regulator of metabolism, differentiation, and cell growth. *J Biol Chem.* 2001; 276:37731–4

Rosenfeld MG, Glass CK. Coregulator codes of transcriptional regulation by nuclear receptors. *J Biol Chem.* 2001; 276(40): 36865-8

Rosmond R, Chagnon M, Bouchard C. The Pro12Ala PPARgamma2 gene missense mutation is associated with obesity and insulin resistance in Swedish middle-aged men. *Diabetes Metab Res Rev.* 2003; 19(2):159-63

Rotllan N, Llaverías G, Julve J, Jauhiainen M, Calpe-Berdiel L, Hernández C, Simó R, Blanco-Vaca F, Escolà-Gil JC. Differential effects of gemfibrozil and fenofibrate on reverse cholesterol *Biochim Biophys Acta.* 2011; 1811(2):104-10

Rotondo D, Davidson J. Prostaglandin and PPAR control of immune cell function. *Immunology*. 2002; 105:20-2

Rusyn I, Asakura S, Pachkowski B, Bradford BU, Denissenko MF, Peters JM, Holland SM, Reddy JK, Cunningham ML, Swenberg JA. Expression of base excision DNA repair genes is a sensitive biomarker for in vivo detection of chemical-induced chronic oxidative stress: identification of the molecular source of radicals responsible for DNA damage by peroxisome proliferators. *Cancer Res*. 2004; 64:1050-7

Sacks FM. After the Fenofibrate Intervention and Event Lowering in Diabetes (FIELD) study: implications for fenofibrate. *Am J Cardiol.* 2008; 102(12A):34L-40L

Saha S, New LS, Ho HK, Chui WK, Chan EC. Investigation of the role of the thiazolidinedione ring of troglitazone in inducing hepatotoxicity. *Toxicol Lett.* 2010; 192(2):141-9

Saidi SA, Holland CM, Charnock-Jones DS, SmithSK. In vitro and in vivo effects of the PPAR-alpha agonists fenofibrate and retinoic acid in endometrial cancer. *Mol Cancer*. 2006; 5:13

Sarraf P, Mueller E, Jones D, King FJ, DeAngelo DJ, Partridge JB, Holden SA, Chen LB, Singer S, Fletcher C, Spiegelman BM. Differentiation and reversal of malignant changes in colon cancer through PPARgamma. *Natl. Med.* 1998; 4, 1046–52

Sato H, Ishihara S, Kawashima K, Moriyama N, Suetsugu H, Kazumori H, Okuyama T, Rumi MA, Fukuda R, Nagasue N, Kinoshita Y. Expression of peroxisome proliferator-activated receptor (PPAR)gamma in gastric cancer and inhibitory effects of PPARgamma agonists. *Br J Cancer.* 2000; 83(10):1394-400

Savoye G. Ulcerative colitis and PPARγ ligand. Is cardiac toxicity on the other side of the coin? *AM J Gastroenterol*. 2008; 103:1571

Schmuth M, Haqq CM, Cairns WJ, Holder JC, Dorsam S, Chang S, Lau P, Fowler AJ, Chuang G, Moser AH, Brown BE, Mao-Qiang M, Uchida Y, Schoonjans K, Auwerx J, Chambon P, Willson TM, Elias PM, Feingold KR. Peroxisome proliferator-activated receptor (PPAR)- beta/delta stimulates differentiation and lipid accumulation in keratinocytes. *J. Invest. Dermatol.* 2004; 122:971–83

Schoonjans K, Martin G, Staels B, Auwerx J. Peroxisome proliferator-activated receptors *Curr Opin Lipidol.* 1997; 8(3):159-66

Semple RK, Chatterjee VK, O'Rahilly S. PPAR gamma and human metabolic disease. *J Clin Invest.* 2006; 116(3):581-9

Shalev A, Siegrist-Kaiser CA, Yen PM, Wahli W, Burger AG, Chin WW, Meier CA. The peroxisome proliferator-activated receptor alpha is a phosphoprotein: regulation by insulin *Endocrinology*. 1996; 137(10):4499-502

Shao D, Rangwala SM, Bailey ST, Krakow SL, Reginato MJ, Lazar MA. Interdomain communication regulating ligand binding by PPAR-gamma. *Nature*. 1998; 396(6709):377-80

Sher T, Yi H-F, McBride OW, Gonzales FJ. cDNA cloning, chromosomal mapping, and functional characterization of the human peroxisome proliferator activated receptor. *Biochemistry*. 1993; 32:5598-604

Shu H, Wong B, Zhou G, Li Y, Berger J, Woods JW, Wright SD, Cai T-C. Activation of PPAR α or γ reduces secretion of matrix metalloproteinase 9 but not interleukin 8 from human monocytic THP-1 cells. *Biochem Biophys Res Commun*. 2000; 267:345–9

Shureiqi I, Jiang W, Zuo X, Wu Y, Stimmel JB, Leesnitzer LM, Morris JS, Fan HZ, Fischer SM, Lippman SM. The 15-lipoxygenase-1 product 13-S-hydroxyoctadecadienoic acid *Proc Natl Acad Sci U S A*. 2003; 100(17):9968-73

Souissi IJ, Billiet L, Cuaz-Pérolin C, Slimane MN, Rouis M. Matrix metalloproteinase *Exp Cell Res*. 2008; 314(18):3405-14

Spiegelman BM. PPAR-gamma: adipogenic regulator and thiazolidinedione receptor. *Diabetes*. 1998; 47(4):507-14

Staels B, Fruchart JC. Therapeutic roles of peroxisome proliferatoractivated receptor agonists. *Diabetes*. 2005; 54:2460–70

Stephen RL, Gustafsson MC, Jarvis M, Tatoud R, Marshall BR, Knight D, Ehrenborg E, Harris AL, Wolf CR, Palmer CN. Activation of peroxisome proliferator-activated receptor delta stimulates the proliferation of human breast and prostate cancer cell lines. *Cancer Res*. 2004; 64:3162–70

Straus DS, Glass CK. Cyclopentenone prostaglandins: new insights on biological activities and cellular targets. *Med Res Rev*. 2001; 21:185-210

Suh N, Wang Y, Williams CR, Risingsong R, Gilmer T, Willson TM, Sporn MB. A new ligand *Cancer Res*. 1999; 59(22):5671-3

Sundvold H, Lien S. Identification of novel peroxisome proliferator-activated receptor (PPAR) gamma promoter in man and transactivation by the nuclear receptor RORalpha1. *Biochem Biophys Res Commun*. 2001; 287: 383–90

Suter SL, Nolan JJ, Wallace P, Gumbiner B, Olefsky JM. Metabolic *Diabetes Care*. 1992; 15(2):193-203

Suzuki T, Hayashi S, Miki Y, Nakamura Y, Moriya T, Sugawara A, Ishida T, Ohuchi N, Sasano H. Peroxisome proliferator-activated receptor gamma in human breast carcinoma: a modulator of estrogenic actions. *Endocr Relat Cancer.* 2006; 13:233-50

Takano H, Komuro I. Peroxisome Proliferator-Activated Receptor γ and Cardiovascular Diseases. *Cir J.* 2009; 73:214-20

Tanaka K, Smith PF, Stromberg PC, Eydelloth RS, Herold EG, Grossman SJ, Frank JD, Hertzog PR, Soper KA, Keenan KP. Studies of early hepatocellular proliferation and peroxisomal proliferation in Sprague-Dawley rats treated with tumorigenic doses of clofibrate. *Toxicol Appl Pharmacol.* 1992; 116:71-7

Tavares V, Hirata RD, Rodrigues AC, Monte O, Salles JE, Scalissi N, Speranza AC, Hirata MH. Association between Pro12Ala polymorphism of the PPAR-gamma2 gene and insulin sensitivity in Brazilian patients with type-2 diabetes mellitus. *Diabetes Obes Metab.* 2005; 7(5):605-11

Teissier E, Nohara A, Chinetti G, Paumelle R, Cariou B, Fruchart JC, Brandes RP, Shah A, Staels B. Peroxisome proliferator-activated receptor alpha induces NADPH oxidase activity in macrophages, leading to the generation of LDL with PPAR-alpha activation properties. *Circ Res.* 2004; 95:1174–82

Teunissen BE, Smeets PJ, Willemsen PH, De Windt LJ, Van der Vusse GJ, Van Bilsen M. Activation of PPARdelta inhibits cardiac fibroblast proliferation and the transdifferentiation into myofibroblasts. *Cardiovasc Res.* 2007; 75:519–29

Theocharis S, Kanelli H, Politi E, Margeli A, Karkandaris C, Philippides T, Koutselinis A. Expression of peroxisome proliferator-activated receptor-gamma in non-small cell lung carcinoma: correlation with histological type and grade. *Lung Cancer.* 2002; 36:249–55

Theodoropoulos G, Papaconstantinou I, Felekouras E, Nikiteas N, Karakitsos P, Panoussopoulos D, Lazaris ACh, Patsouris E, Bramis J, Gazouli M. Relation between common polymorphisms in genes related to inflammatory response and colorectal cancer. *World J Gastroenterol.* 2006; 12(31):5037-43

Thevis M, Möller I, Thomas A, Beuck S, Rodchenkov G, Bornatsch W, Geyer H, Schänzer W. Characterization of two major urinary metabolites of the PPARdelta-agonist GW1516 and implementation of the drug in routine doping controls. *Anal Bioanal Chem.* 2010; 396(7):2479-91

Tönjes A, Stumvoll M. The role of the Pro12Ala polymorphism in peroxisome proliferator-activated receptor gamma in diabetes risk. *Curr Opin Clin Nutr Metab Care.* 2007; 10(4):410-4

Tontonoz P, Hu E, Spiegelman BM. Stimulation of adipogenesis in fibroblasts by PPAR gamma 2, a lipid-activated transcription factor. *Cell.* 1994; 79:1147–56

Tontonoz P, Nagy L, Alvarez JGA, Thomazy VA, Evans RM. PPARγ promotes monocyte/macrophage differentiation and uptake of oxidized LDL. *Cell. 1998; 93:241-52*

Tordjman K, Bernal-Mizrachi C, Zemany L, Weng S, Feng C, Zhang F, Leone TC, Coleman T, Kelly DP, Semenkovich CF. PPARalpha deficiency reduces insulin resistance and atherosclerosis in apoE null mice. *J Clin Invest.* 2001; 107:1025–34

Tschritter O, Fritsche A, Stefan N, Haap M, Thamer C, Bachmann O, Dahl D, Maerker E, Teigeler A, Machicao F, Häring H, Stumvoll M. Increased insulin *Metabolism.* 2003; 52(6):778-83

Valve R, Sivenius K, Miettinen R, Pihlajamaki J, Rissanen A, Deeb SS, Auwerx J, Uusitupa M, Laakso M. Two polymorphisms in the peroxisome proliferator-activated receptor-gamma gene are associated with severe overweight among obese women. *J Clin Endocrinol Metab.* 1999; 84:3708–12

Vogels N, Mariman EC, Bouwman FG, Kester AD, Diepvens K, Westerterp-Plantenga MS. (2005): Relation of weight maintenance and dietary restraint to peroxisome proliferator-activated receptor γ2, glucocorticoid receptor, and ciliary neurotrophic factor polymorphisms. *Am J Clin Nutrition.* 2005; 82(4): 740–6

Wang XL, Oosterhof J, Duarte N. Peroxisome proliferator-activated receptor gamma C161→T polymorphism and coronary artery disease. *Cardiovasc Res.* 1999; 44:588–94

Wang P, Anderson PO, Chen S, Paulsson KM, Sjogren H-O, Li S. Inhibition of the transcription factors AP-1 and NF-κB in CD4 T cells by peroxisome proliferator-activated receptor-γ ligands. *Int Immunopharmacol.* 2001; 1:803-12

Wahli W, Braissant O, Desvergne B. Peroxisome proliferator activated receptors: transcriptional regulators of adipogenesis, lipid metabolism and more. *Chem Biol.* 1995; 2:261-6

Wang X, Kilgore MW. Signal cross-talk between estrogen receptor alpha and beta and the peroxisome proliferator-activated receptor gamma1 in MDA-MB-231 and MCF-7 breast cancer cells. *Mol Cell Endocrino.* 2002; 194:123-33

Warren JR, Simmon VF, Reddy JK. Properties of hypolipidemic peroxisome proliferators in the lymphocyte [3H]thymidine and Salmonella mutagenesis assays. *Cancer Res.* 1980; 40(1):36-41

Weigel N.L. Steroid hormone receptors and their regulation by phosphorylation. *Biochem. J.* 1996; 319: 657-67

Woods CG, Kosyk O, Bradford BU, Ross PK, Burns AM, Cunningham ML, Qu P, Ibrahim JG, Rusyn I. Time course investigation of PPARalpha- and Kupffer cell-dependent effects of WY-14,643 in mouse liver *Toxicol Appl Pharmacol.* 2007; 225(3):267-77

Wu KK, Liou JY. Cyclooxygenase inhibitors induce colon *Methods Mol Biol.* 2009; 512:295-307

Yaacob NS, Mohd. Darus H, Norazmi MN. Modulation of cell growth and PPARγ expression in human colorectal cancer cell lines by ciglitazone. *Exp Toxicol Pathol.* 2008; 60(6):505-12

Yaacob NS, Mohd. Ariffin K, Norazmi MN. Differential transcriptional expression of PPARα, PPARγ1 and PPARγ2 in the peritoneal macrophages and T cell subsets of non obese diabetic mice. *J Clin Immunol.* 2009; 29:595-602

Yaacob NS, Goh KSK, Norazmi MN. Male and female NOD mice differentially express peroxisome proliferator-activated receptors and pathogenic cytokines. *Exp Toxicol Pathol.* 2010; [Epub ahead of print]

Yang XY, Wang LH, Chen T, Hodge DR, Resau JH, DaSilva L, Farrar WH. Activation of human T lymphocytes is inhibited by peroxisome proliferator activated receptor-□ (PPAR gamma) agonists *J Biol Chem.* 2000; 275:4541-4

Yang L, Zhou ZG, Zheng XL, Wang L, Yu YY, Zhou B, Gu J, Li Y. RNA Interference against peroxisome proliferator-activated receptor δ gene promotes proliferation of human colorectal cancer cells. *Dis Colon Rectum.* 2008; 51:318–28

Yen CJ, Beamer BA, Negri C, Silver K, Brown KA, Yarnall DP, Burns DK, Roth J, Shuldiner AR. Molecular scanning of the human peroxisome proliferator activated receptor gamma (hPPAR gamma) gene in diabetic Caucasians: identification of a Pro12Ala PPAR gamma 2 missense mutation. *Biochem Biophys Res Commun.* 1997; 241(2):270-4

Yokoyama Y, Xin B, Shigeto T, Umemoto M, Kasai-Sakamoto A, Futagami M, Tsuchida S, Al-Mulla F, Mizunuma H. Clofibric acid *Mol Cancer Ther.* 2007; 6(4):1379-86

Youssef J, Badr M. Role of peroxisome proliferator-activated receptors in inflammation control. *J Biomed Biotechnol.* 2004; 2004:156-66

Zandbergen F, Plutzky J. PPARalpha in atherosclerosis *Biochim Biophys Acta.* 2007; 1771(8):972-82

Zelcer N, Tontonoz P. SUMOylation and PPARgamma: wrestling with inflammatory signaling. *Cell Metab.* 2005; 2:273–5

Zhang X, Young HA. PPAR and immune system *Int Immunopharmacol.* 2002; 2(8):1029-44

Zhu Y, Qi C, Korenberg JR, Chen XN, Noya D, Rao MS, Reddy JK. Structural organization of mouse peroxisome proliferator-activated receptor gamma (mPPAR gamma) gene: alternative promoter use and different splicing yield two mPPAR gamma isoforms. *Proc Natl Acad Sci U S A. 1995;* 92(17):7921-5

Zuo X, Wu Y, Morris JS, Stimmel JB, Leesnitzer LM, Fischer SM, Lippman SM, Shureiqi I. Oxidative metabolism of linoleic acid modulates PPAR-beta/delta suppression of PPAR-gamma activity *Oncogene*. 2006; 25(8):1225–41

Zuo X, Peng Z, Moussalli MJ, Morris JS, Broaddus RR, Fischer SM, Shureiqi I. Targeted genetic disruption of peroxisome proliferator-activated receptor-delta and colonic tumorigenesis. *J Natl Cancer Inst.* 2009; 101(10):762-7

INDEX

#

A

D

H

I

K

L

M

N

O

P

R

U

V

W

X

Z